T0143060

MULTICRITERIA ENVIRONMENTAL ASSESSMENT

Multicriteria Environmental Assessment

A Practical Guide

by

Nolberto Munier
Ottawa, Ontario, Canada

KLUWER ACADEMIC PUBLISHERS
DORDRECHT / BOSTON / LONDON

A C.I.P. Catalogue record for this book is available from the Library of Congress.

ISBN 1-4020-2089-9 (PB)
ISBN 1-4020-2088-0 (HB)
ISBN 1-4020-2090-2 (e-book)

Published by Kluwer Academic Publishers,
P.O. Box 17, 3300 AA Dordrecht, The Netherlands.

Sold and distributed in North, Central and South America
by Kluwer Academic Publishers,
101 Philip Drive, Norwell, MA 02061, U.S.A.

In all other countries, sold and distributed
by Kluwer Academic Publishers,
P.O. Box 322, 3300 AH Dordrecht, The Netherlands.

Printed on acid-free paper

Printed in the Netherlands.

TABLE OF CONTENTS

PART I : GENERAL BACKGROUND 1

CHAPTER 1 - GENERAL INFORMATION 3
1.1 The purpose of this book 3
1.2 Structure 3
1.3 Environmental Impact Assessment (EIA) 8
1.4 Environmental Impact Statement (EIS) 9
1.5 Strategic Environmental Assessment (SEA) 11
1.6 Environmental and Socioeconomic Impact Assessment (ESIA) 12
1.7 The necessity of developing an Environmental Impact Assessment (EIA) for most projects 13
1.8 What is economic development? 13
1.9 What is sustainability? 14
1.10 The link between economic development and sustainable environment 15
1.11 The carrying capacity of the environment 17
1.12 Social, economic and environmental interaction in sustainable development 17
1.13 The EIA problem 18
1.14 The decision-making process 20
1.15 The EIA report 21
Internet references for Chapter 1 23

CHAPTER 2 - PROJECTS AND IMPACTS 27
2.1 Defining the objectives of the study 27
2.1.1 Projects, alternatives or options 27
2.1.2 Why are there alternatives to a single project? 29
2.1.3 Inventory of alternatives or options 30
2.1.4 Identification of criteria to compare alternatives 30
2.1.5 Projects' characteristics 31
2.1.5.1 Selection of alternatives 31
2.1.5.2 Public opinion 31
2.1.5.3 What are the problems for appraisal? 32
2.1.5.4 The baseline 33
2.2 Impacts caused by projects 34
2.2.1 Impact definition 34
2.2.2 Effects caused by projects 34
2.2.2.1 Positive or adverse 34
2.2.2.2 Primary, secondary, tertiary 35
2.2.2.3 Measurable or indeterminate. 36
2.2.2.4 Apparent impacts 36
2.2.2.5 Cumulative 37

2.2.2.6 Able to be mitigated or not in a greater or lesser degree 37
2.2.2.7 Residual impact 37
2.2.2.8 Spatially related 38
2.2.2.9 Temporal related 38
2.2.2.10 Irreversible or reversible 39
2.2.2.11 Likelihood of the impacts 39
2.2.2.12 Unexpected impacts 39
2.2.2.13 Risk effects 40
2.2.2.14 Residual effects 41
2.2.2.15 Population impact 41
2.2.2.16 Interaction between impacts 42
2.3 Conclusion on impacts 42
Internet references for Chapter 2 44

CHAPTER 3 - CRITERIA 47
3.1 Criteria, Attributes or Components 47
3.1.1 Technical criteria 47
3.1.2 Environmental criteria 48
3.1.3 Safety criteria 49
3.1.4 Social criteria 50
3.1.5 Economic criteria 50
3.1.6 Construction criteria 50
3.1.7 Spatial criteria 51
3.1.8 Temporal criteria 51
3.1.9 Diverse areas covered by criteria 51
3.1.10 Criteria weight 52
3.1.11 Criteria thresholds 53
3.1.12 Threshold standards 54
3.1.13 Criteria relationships 56
3.1.14 Assessment of significance 56
3.1.15 Criteria selection 56
3.2 Factor Analysis (FA) 57
3.3 Steps for Environmental Impact Assessment 59
3.3.1 Screening... 59
3.3.2 Scoping..... 59
3.3.3 Remediation 60
3.3.4 Impact assessment 60
3.3.5 Impact mitigation 61
3.3.6 Public participation 61
Internet references for Chapter 3 64

PART II: ELEMENTS OF ENVIRONMENTAL APPRAISAL 67

CHAPTER 4 - INFORMATION FOR EIA 69
4.1 Defining the problem 69
4.2 Available tools for project appraisal 69
4.3 Tools for impact identification 70
4.3.1 Checklists... 70
4.3.1.1 Case study - Complex for energy generation and transmission 70
4.3.2 Network analysis 76
4.3.2.1 Case study - High speed train project 77
4.3.3 Leopold's matrix. 83
4.3.3.1 Case study - Construction of a water reservoir 84
4.3.4 Flow diagrams 87
4.3.5 Battelle environmental evaluation system 87
4.3.6 McHarg system - Overlays 87
4.3.7 The Delphi method 90
4.3.8 Dose-response functions 92
4.3.9 Stepped or chained matrices 94
4.3.10 Comparison of techniques 95
Internet references for Chapter 4 97

CHAPTER 5 - TECHNIQUES FOR ENVIRONMENTAL APPRAISAL ... 99
5.1 Geographic Information System (GIS) 99
5.1.1 Fundamentals 99
5.1.2 Case study --- Road construction near archaeological ruins 102
5.2 Contingent Valuation (CV) 104
5.2.1 Case study --- Visitors in an environmentally sensitive site 105
5.3 Cost-Benefit Analysis (CBA) 106
5.3.1 Hedonic pricing 109
5.3.2 Case study -- Sewage conversion in a tourist resort 111
5.4 Cost-Effectiveness Analysis (CEA) 114
5.4.1 Case study --- Selection of exhaust filtering equipment 115
5.5 Input-Output analysis (IO) 117
5.5.1 Case study --- Contamination produced by the automobile manufacturing industry 118
5.6 Life Cycle Analysis (LCA) 122
5.6.1 Case study - Construction of a copper concentration plant 125
5.6.2 Case study --- Determination of contamination produced by a metallurgical mining process 127
5.7 Multicriteria Analysis (MCA) 132
5.7.1 Dollar value appraisal 134

5.7.1.1 Case study --- Selection of different urban projects..........134
5.7.2 Dollar value with monetary restrictions 139
5.7.3 Environmental damage appraisal...139
5.7.3.1 Case study - Selection of a chemical process to minimize environmental costs 140
5.8 Analytical Hierarchy Process (AHP)...145
5.8.1 Case study --- Selection of renewable energy sources 146
5.9 Mathematical Programming (MP)...156
5.9.1 Case study --- Construction of an oil pipeline............................160
5.10 Actual examples of multicriteria analysis 167
Example 1: Selection of waste incinerators (several countries)..........168
Example 2: Selection of a sewage system (Norway)...........................168
Example 3: Options to alleviate traffic congestion (The Netherlands) 169
Example 4: Construction of an overhead power line (South Africa) ... 169
Example 5: Routing for an oil pipeline 170
Example 6: Construction of an LPG pipeline (India)..........................170
Example 7: Ranking medium hydro-electric projects (Nepal).............171
Example 8: Policy for energy generation (Canada).............................171
5.11 Risk Analysis (RA)..172
5.12 Techniques comparison..175
5.13 Strategic Environmental Assessment 177
5.13.1 Case study-- Policy implementation for paper recycling...........177

Internet references for Chapter 5 187

PART III: EIA IN THE URBAN SCENARIO 195

CHAPTER 6 - EIA AND URBANIZATION 197
6.1 People's participation..197
6.2 EIA and the city...198
6.2.1 Case study --- Selection of urban intersections 200
6.3 Urban and regional projects...205
6.4 The execution of joint projects..206
6.5 Sustainable impact...207
6.5.1 Case study --- Plan of new dwellings for low-income people......207
6.6 Indicators...209
6.7 Indicators selection..212
6.8 Measuring the quantity of information provided by indicators...........213
6.8.1 Case study --- Selection of urban sustainable indicators.............214
6.8.2 Case study --- Selection of urban indicators and projects 224
6.8.3 Case study --- Slum upgrading 225
6.9 Monitoring..231
Internet references for Chapter 6 232

APPENDIX : MATHEMATICAL BACKGROUND TO UNDERSTAND THE FUNDAMENTALS OF MODERN TECHNIQUES FOR PROJECT APPRAISAL 235
A.1 Introduction 237
A.1.1 Regression analysis 237
A.1.1.1 Example --- Analysis of an urban transportation corridor 238
A.1.2 Correlation analysis 243
A.1.3 Matrices 244
A.1.4 Linking matrices and correlation 245
A.1.5 Linear transformation of a unit circle 247
A.1.6 Relationship between the ellipse and the correlation matrix 248
A.2 Mathematical foundation of the Analytical Hierarchy Process 251
A.2.1 Example --- Analysis of a project with three criteria 251
A.3 Foundation of Mathematical Programming 253
A.3.1 Example --- Analysis of a project that calls for the construction of an oil pipeline, with four alternatives or options 253
A.3.2 Graphic example 256
Project: Improvement of an urban highway 256
A. 4 Mathematical foundation of input/output analysis 261
A.4.1 Example - Railway industry 261
A.5 Statistical notions of Factor Analysis 265
A.6 The notion of opportunity cost 266
A.7 The notion of shadow prices 267
A.8 The notion of sensitivity analysis 268
Internet references for Appendix 268
Cited references on the Internet by industries 269

ROAD MAP FOR ENVIRONMENTAL IMPACT ASSESSMENT 277

Flowchart and road map for Environmental Impact Assessment 279

GLOSSARY 301

BIBLIOGRAPHY 305

INDEX 309

PART I : GENERAL BACKGROUND

CHAPTER 1 - GENERAL INFORMATION

1.1 The purpose of this book

This book is oriented to people who are charged with the task of preparing an EIA for a certain project or projects, and it intends to be a practical guide for a practitioner. It is also addressed to stakeholders, decision makers and other interested parties.

It follows a natural sequence of events and actions conducted to facilitate the purpose of EIA, identifying alternatives or projects that cause minimum environmental impact. In this case, the expression "environmental impact" involves all classes of impacts and scenarios, i.e., is not restricted to the environment.

This concept is important and for this reason we reproduce here the definition from UNEP Training Resource Manual, where it says that "the term "environment" includes:

- Human health and safety;
- Flora, fauna, ecosystems and biodiversity;
- Soil, water, air, climate and landscape;
- Use of land, natural resources and raw materials;
- Protected areas and sites of special significance;
- Heritage, recreation and amenity assets;
- Livelihood, lifestyle and well being of affected communities.

1.2 Structure

The book has been structured in the following manner:

In Chapter I definitions are given regarding the different concepts involved in the environment assessment procedure, and a key question is answered related with the necessity of developing an EIA for most projects.

Then the concept of projects or alternatives is introduced together with the fundamental notion of criteria. Most projects are developed to increase economic development by generating employment, improving the balance of payments, increasing the standard of living, etc., so this important concept is defined and commented discussed.

As a counterpart, economic development also means some damage to the environment that can be reversible or irreversible, which can deplete resources or modify their existence for years to come, therefore the notion of sustainable development and its application is of paramount importance, and the link between these two important concepts is then analyzed.

Sustainable development is closely related with social behavior as well as with public participation, and this last thought is fundamental for the project to have the support of individuals. Besides, a capital concept linked with sustainability is the carrying capacity of the environment, a very important factor that needs to be seriously considered.

Once the problem has been defined, its main components are briefly analyzed; in other words once we have immersed ourselves in the "theatre", the different actors and components of this "play" are introduced.

The "actors" are the people likely to be affected by the project, their representatives, stakeholders and the decision makers. The components of the "play" are:

- The listing of projects or alternatives to be evaluated;
- The framework where the analysis will be conducted;
- The different criteria used to appraise these projects;
- The values linking each criteria with the corresponding project;
- The threshold values given to each criterion;
- The weight assigned to criteria and perhaps to projects;
- The significance of impacts.

This is then the moment where it is necessary to make a plan, to prepare some sort of road map to outline the course of action, including establishing of the objective of the study. For this purpose the different components of the EIA are outlined. Finally, in this chapter the fundamental decision-making process is reviewed, incorporating some sort of a feedback mechanism from affected and interested people.

Chapters 2 and 3 are devoted to the analysis of the components to be considered in an EIA, that is the projects, alternatives or options, the impacts they produce and the criteria employed for their appraisal. Thus, in specific projects it is discussed why it is necessary to take into account different alternatives even for a single project, and also the very important concept of baseline is explored. The different kinds of impacts are investigated together with the diverse categories of criteria that are usually employed in an EIA. Also, the steps for executing an EIA are studied, explaining in each case their meaning and scope.

Part II of the book begins with inspection of the elements used for environmental appraisal. To this extent, Chapter 4 deals with collecting information for the EIA, and as a consequence several techniques are discussed with case studies to exemplify some of them. Once the data is collected, the subsequent task is to use that information to make the appraisal of a project or the selection of its alternatives, or even their ranking. From this point of view seven techniques are detailed in Chapter 5, they are:

- Geographic Information Systems (GIS);
- Contingent Valuation (CV);
- Cost-Benefit analysis (CBA);
- Cost-Effectiveness analysis (CEA);
- Input-Output analysis (IO);
- Life Cycle analysis (LCA);
- Multicriteria analysis (MCA).

Each of these techniques is illustrated with a case study, as follows:

- The construction of a road with possible adverse effects in a nearby archaeological site, for GIS;
- A plan to introduce a fee for people visiting a natural cave in order to preserve the site, for CV;
- A study to connect dwellings to a water treatment plant, to avoid contamination in a tourism area, for CBA;
- A study to select filtering equipment for gases from an industrial furnace, for CEA;
- Calculation of contamination produced by the automobile manufacturing industry, for IO;
- Calculation of the amount of gases released in a metallurgical mining process, for LCA;
- Selection of different urban projects for MCA;
- Selecting of renewable energy sources for MCA (using AHP);
- Construction of an oil pipeline, for MCA (using MP).

Preference is given to Multicriteria Analysis because it is believed to be the most complete technique and also because it is nowadays the most preferred tool in many countries. For this technique, five different approaches are illustrated, with case examples for each one, three of them mentioned above. This section concludes with Risk Analysis, which is also discussed, and with a comparison of techniques. This is not a detailed comparison but a very short one, pointing out the basic strengths and weaknesses of each methodology. The general consensus is that there is no " superior" technique, but that as a matter of fact all of them are complementary to each other.

Strategic Environmental Assessment is also briefly reviewed and a case study proposed dealing with the implementation of a government policy to recycle paper. As a final but very important part of Part II, eight actual examples, corresponding to the same number of EIA studies by several practitioners around the world, are very briefly discussed.

These case studies include different issues such as: solid waste, sewage, highways, energy, oil and gas pipelines, hydro-projects, and energy generation and transmission. In each case information is available for the reader to consult the whole paper either via the Internet or in the Bibliography section. It is believed that these case studies will provide the reader with a very valuable source of information and will illustrate different modus operandi in a variety of issues.

In Part III the text enters the field of EIA when applied to the urban context. Because it is here that public participation will be at its most earnest, and also considering that a city is usually the recipient of many projects, the relationship between the project and the people is analyzed, as well as the proposal of measures for people to participate.

Two case studies are offered here, one of which is a plan related with the selection of urban intersections. It has been chosen with the main purpose of demonstrating a feature of many projects, which is their inter-dependency, as well as the characteristic of others which stand alone or are exclusive.

The other case example deals with an urban problem faced in many countries around the world, and that is how to improve and integrate the urban poor in the city fabric. The existence of slums pose serious environmental and social problems, and their relocation, which can be considered a project, also can have severe environmental impacts and consequences.

A crucial theme is analyzed here, and is related with urban indicators. Many of these indicators can be used as criteria in non-urban projects, but a city has its own problems that call for their urban pointers. Experience shows that there could be hundreds of indicators, and then multicriteria analysis is applied to determine a suitable final list of manageable indicators, and subjected to many restrictions, constraints or criteria. The second case study refers to a selection of either cities in a country or areas within the city where there is an opportunity to upgrade slums through the provision of basic infrastructure services, which also pose environmental as well as sustainability problems.

Part III finishes with the monitoring activity, that is with the quantitative and qualitative control of the efficiency of measures taken to keep the effects of impacts and especially the residual impacts under control.

The Appendix offers a brief refresher course on especially important mathematical subjects that are important for understanding the foundations of the different tools and techniques discussed in Chapters 4, 5 and 6. In each of these chapters the relevant technique is explained and case studies proposed to demonstrate specific applications. However, no information is given about the mathematics behind each method, and this is the purpose of this Appendix.

Key concepts are covered on regression, correlation, matrices, and eigen-value analysis, as well as linear transformations. There is no mathematical rigor in these comments, just plain language to explain them, and because of that, the explanations will not conform to that of a mathematician but will illustrate the approach of the practitioner. Also four case studies are offered here for four very important techniques: Regression, Analytical Hierarchy Process, Mathematical Programming and I-O analysis.

A section called *Road Map for Environmental Impact Assessment* provides a flowsheet of the whole process as well as a expanded flowsheet. This document, covering all aspects of the preparation of an EIA Report has been designed to be used as a worksheet, and each topic carries references related to that topic in the book.

The Glossary supplies some short definitions, and again, without mathematical rigor. Finally there are two classes of References: One is linked with Internet sites, so that the reader can access immediately the information he/she wants to consult. The other is a classical Bibliography referring to books and papers of interest.

1.3 Environmental Impact Assessment (EIA)

EIA is a systematic analysis of the potential impacts that a project can produce on the environment, and it tries to find a balance between gains and losses.

This analysis involves:

- Identification of the objective;
- Continues with recognition of the alternatives or projects;
- Goes on with the selection of criteria to be used to gauge the contribution of each project to the achievement of the objective through the established criteria;
- Advances the task of assessing the project impacts and their effects;
- Gathers information and data regarding the effects of the project;
- Makes the selection of alternatives;
- Proposes the monitoring process;
- Establishes the format for the Environmental Impact Assessment;
- Status report.

At the beginning, commencing with the implementation of this discipline, EIA was project specific, working at the project level. This is an early definition and nowadays the term applies to the analysis and assessment of environmental impacts produced by many projects or alternatives options subject to many criteria, i.e., how each project affects the different resources involved in its development.

Today, EIA is the generic name applied to a set of appraisal techniques, however, there is a technique which is called EIA. It is a very important one, because even with its drawbacks and short of providing a decision rule for aggregating environmental impacts, it gives information and data which are common to all other techniques employed.

Probably the first question is if EIA is mandatory for all classes and type of projects. The answer is that in general it is not, but the decision depends on regulations in different countries. Some projects under a certain size do not necessitate an EIA, others need one whatever the size. As a general idea, and not as a limiting example, it can be assumed that:

Projects requiring EIA whatever their size are:

- Portland cement plants;
- Mining and processing of certain minerals;
- Fish processing;
- Oil extraction and processing;
- Waste incinerators.

Projects requiring EIA if they exceed a certain size:

- Urban developments;
- Industrial developments;
- Certain type of chemical factories;
- Waste water treatment plants stations;
- Landfills.

In EIA two stages can be considered:

- Project appraisal. In here the environmental assessment report is reviewed and public consultation takes place.

- Project evaluation. In this stage proceeds the evaluation of the environmental assessment report, and the corresponding mitigation plan.

The concepts of assessment, quantification, evaluation and monitoring in EIA are now introduced. *What is to be assessed?* The impact of actions to be generated by the project. *What is to be quantified?* The magnitude of the impact. *What is to be evaluated?* Here we have to put ourselves in the future when we will be in a condition to analyze the effects that those actions have produced in human health, well-being and in the environment. *What is to be monitored?* Compliance of the measures proposed to mitigate the project effects as well as verification that the residual effects are within the limits imposed.

1.4 Environmental Impact Statement (EIS)

This statement is mandatory for projects with impacts that have a significant effect on the environment. It shows the project, its alternatives, scale, approaches, potential impacts and remediation measures, and likely residual impacts. It is also the proposed monitoring procedure.

As an example, projects that fall into this category are:

- *Large farming operations* involving livestock such as the swine, beef, and poultry industries. One of the main problems here is the generation of manure, which not only contaminates the ground where it falls but because it is transported by rain to rivers, polluting water sources. One way to take care of the problem in a hog farm for instance is to collect the manure from the pig pens through grate floors. Then it is mixed with water and sent to a septic tank.
  When this is filled the liquid manure is piped to an outdoor tank from where it is sold as a fertilizer. The liquid manure has potassium, nitrogen and phosphorus and it is said to cut fertilizer bills by 25% when it is used on different crops.This is an example of waste management which has to be detailed in the EIS.

- *Treatment of spent fuel rods from nuclear reactors.* This is a very important problem that has not found yet a satisfactory solution. The United States for instance has a plan to bury this waste in a geological repository in the Yucca Mountains in the State of Nevada, although another solution could be perhaps the conversion of this waste into a more stable product or be used for other purposes.
- *Forest logging* in some sensitive areas around the world falls also in this category. For instance, and as unbelievable as it sounds, some decades ago there was a project to use the head of the Iguazú Falls in South America to generate electricity through the construction of a dam upstream. Of course it would have destroyed one of the most spectacular water falls in the world, and fortunately it did not prosper.

- In some rivers there are several hydro-electric dams constructed along its length. In many cases this creates a problem for *salmon migration* through the dams and endangers juvenile fish survival during their trip to the ocean.

1.5 Strategic Environmental Assessment (SEA)

SEA can be defined as the formalized, systematic and comprehensive process of evaluating the environmental impacts of a policy, plan or program and its alternatives, the preparation of a written report on the findings, and the use of the findings in publicly accountable decision making (see Bibliography - Therivel *et al.* 1992). As a consequence, SEA can be defined as a process whose output is information used to sustain decision-making on issues related with strategic planning. Its purpose is similar to EIA's and most of the techniques are also the same, but they differ in their utilization. EIA is applied to the evaluation and selection of alternative forms of projects, when all significant environmental impacts are considered. The key words in this sentence are "alternatives" or "projects". Because of its scope, EIA works at project level.

SEA is applied to the evaluation and selection of policies when all significant environmental impacts are considered. The key word here is "policies". This difference is not semantic, for the dimension and the effects of both techniques differ.

Because EIA applies to alternative forms of a project or to different projects with several alternatives each, their influence in the economic and social fabric of a region could be very important, but it could not be so important at the national level.

If a plant manufacturing gear boxes for the automobile industry and employing 1,300 persons closes for whatever reasons, the closing will probably have a large impact in the city where it is located. The smaller the city the larger the social and economic impacts. However, at a national level it can be very small, when gauged for instance against the National Gross Product.

Now, if we consider SEA and because it refers to policies, the effects of implementing these policies will have a great incidence in the population because it can affect the whole country. An example could be a policy to reduce emissions in accordance with the Kyoto Protocol. Besides, the effects of a policy can last for decades, which is another difference with the EIA model. This text is mainly oriented to EIA, however some examples are given using techniques for SEA.

1.6 Environmental and Socioeconomic Impact Assessment (ESIA)

In reality this document should be a part of the EIA report but its importance is so great that it merits a separate study. Many reports mention the impacts produced by the projects and to some extent the mitigation measures; however, especially in the social sector this information is often missing, since there are generally alternatives or options for a project, but many times there are none for the social aspect. For instance, in the case of the construction of a dam, where the formed lake will flood a nearby town, there is for sure a provision that a new town will be constructed in another place and its inhabitants relocated, but usually there are no other alternatives regarding this part of the project.

Have the people been consulted about other options? Are they happy with the monetary compensation offered for the flooding of their arable land? Does the relocation area have the same characteristics, so as to enable people to continue with their activities?

Is there any provision for the people to buy electricity at a lower price considering the upsetting of their way of life? If the people are relocated in a nearby existing town, what is the quality of the services they can expect regarding education, healthcare, utilities, etc.?

Regarding the environment, similar questions arise. For instance: Is there a waste management plan in place to handle wastes from the project during the construction and operation stages? Is there a procedure in place when involuntary spills or fuel impregnate the ground? If so, what is the remediation procedure?

In some projects conventional equipment is used, but in others the heavy equipment was built only for that purpose. This is the case for instance for a tunnel boring machine whose most important component --- the rotating head ---, as well as its structure were built for a specific diameter, so unless there is another construction with the same geometric characteristics it is most probable that this equipment will be sold for scrap. However, sometimes the scrap value does not cover the scrapping costs and the equipment is abandoned. As a dramatic example remember that when the excavation of the Channel Tunnel was completed, the English Tunnel Boring Machine was made to excavate its own tunnel, off the main tunnel, and then was buried there.

Many decades after streetcars were banned in many cities around the world, their steel tracks can still be seen embedded in the pavement. Considering this fact, a plan to remove these tracks should have been devised --- even if it was not economical ---, in order not to leave them as junk.

In conclusion, it is necessary to prepare an ESIA that details all the measures to be taken in order to properly address such issues.

1.7 The necessity of developing an Environmental Impact Assessment (EIA) for most projects

Mankind needs to develop projects for political or egotistic reasons (Egypt Pyramids), for military mobilization (The Appian Way in Rome), to show his faith (Temples all around the world), for defense (the Adrian Wall in England), to generate electricity (nuclear plants), for transportation (modern highways), etc. One wonders what effects or consequences any given project will produce?
The Dictionary defines the word "effect" as "something brought about by a cause or agent: a result". This is a good definition for our purposes. For all projects bring some degree of perturbation of the social fabric, because they involve people, induce environmental changes due to the use of non-renewable resources, have an economic impact since monetary resources are used, or cause sustainable loss because a resource is depleted or badly affected. In other words there are always some results, positive and negative, and they need to be evaluated.

1.8 What is economic development?

Many definitions have been proposed for economic development. In essence, it is usually a long term process, that promotes growth, encourages competitiveness, increases employment opportunities and wages, enhances higher education, reduces poverty, diminishes inequalities, etc.

Consider for instance the impact that a new industry such as a petrochemical plant brings to an area. It offers employment, promotes the need for the people to learn a new trade and enhances their ability to find new jobs, builds roads, triggers commerce activities, etc., that is, the spin-offs and multiplier effects of such a project are considerable. Remember that this employment, in turn, generates more work, since people acquire more disposable income and can then buy appliances, cars, build houses, travel, etc., which in turn produces more benefits.

It is obvious that economic development offers a better standard of life when measured in the terms above mentioned; however, most often that improvement places a heavy burden on the environment, because it also produces pollution, depletes non-renewable resources, increases wastes, uses more water, etc.

Considering all of the above, it is necessary to bear in mind that any EIA study has to take into account the potential advantages of a project, or different projects, that aim to improve the quality of life or to generate an economic benefit, but at the same time has to reflect the damages that the environment will suffer because of this economic undertaking. Such progress cannot be stopped since our world needs increasing use of natural resources to build roads and railways for better transportation, to supply clean and abundant drinking water, to provide adequate health care, to build more houses and commercial centers, to manufacture more equipment for better communication, etc.

On the other hand it is necessary to think deeply about available resources, either non-renewable such as minerals and oil, or renewable like land, water, air, forest, etc. The trouble is that even renewable resources are used to such an extent that they can be depleted, such as fish, forests, land, and of course water. So, how can these two opposite aims be harmonized, that is to promote economic development but at the same time to keep resources available for generations to come? This is in reality the main goal of an EIA study, since it has to find an equilibrium, a balance, between gains offered by a project against losses that the society is willing to accept in the quality of their environment.

1.9 What is sustainability?

There are many definitions of this term, but the most accepted nowadays is that created by the World Commission on Environment and Development, and that was included in the so-called Bruntland Report, that defines sustainable development as that ***"which meets the needs of the present population without compromising the ability of future generations to meet their own needs***". (see Bibliography, World Commission on Environment and Development).

A very interesting paper, entitled ***"It is too late to achieve sustainable development, now let us strive for survivable development"***, was presented by Dennis. L. Meadows, in November 1994 at the Eighth Toyota Conference in Japan. http://www.unu.edu/unupress/unupbooks/uu03pe/uu03pe0c.htm (see also Bibliograpphy --Meadows, D.L.). He prepared Table 1.1, showing three different conditions, Sustainable, Critical and Destructive, with values for different indicators.

Table 1.1 **Criteria for Sustainable Development**

Human activity	Sustainable	Critical	Destructive
Population increase	< 0.5% per year	1.0 -1.5 % per year	>2% per year
Economic development	3% < GNP < 5%	8% < GNP <10%	GNP > 10% (over development) GNP < 0% (under development)
Deforestation rate	< 0.1% per year	0.5-1.0 % per year	> 1% per year
Forest coverage	> 30 %	15-20%	< 10 %
Agricultural development	> 0.3% ha / capita	0.15-0.2 ha / capita	0.1 ha / capita
Self-support ratio	> 91 %	60-70 %	< 50 %
Population density	< 50 / ha	100-150 / ha	> 200 / ha
Population of a city	< 0.5 million	> 1 million	> 10 million

Source: Dennis Meadows
Reprinted with kind permission of Dennis Meadows

It is worth noticing how economic development, when it exceeds a certain limits, puts so heavy a pressure on the sustainable resources that it destroys them. It is possible to use sustainability indicators as another type of criteria, especially in multicriteria analysis.

1.10 The link between economic development and sustainable environment

Our situation is something like asking a hungry person to satisfy his/her hunger with the minimum possible amount of food; the requirements are contradictory. Economic development implies construction of roads, laying of oil and gas pipelines, mining of minerals, exploitation of sea resources, paving of streets, use of pesticides, increase of transportation, etc. The effects produced by these projects are:

1. The construction of roads not only consumes non-renewable resources but also produces pollution.
2. The laying of a gas and oil pipeline promotes the use of non-renewable resources.
3. Mining the minerals alters the environment, and produces pollution.
4. The exploitation of a sea creature resource could deplete the species.
5. The paving of streets alters the natural rate for replenishment of aquifers.

6. The use of pesticides and nutrients, modifies the soil composition, and pollutes the water of the rivers.
7. The increase of transportation increments the use of non-renewable resources (fossil fuels) and contaminates a renewable resource (air) sometimes beyond its regeneration limits in a certain period of time (look at the contaminated air of Mexico City, São Paulo, Bangkok, Tokyo, Los Angeles, etc.).

But on top of the damaging effect that some of these economic activities have on the environment, there are others that also challenge the sustainability of our resources. Of course, probably the best example is the consumption of a non-renewable resource such as oil and gas. But what about renewable resources?

For instance we can have a project to build a new housing development for low income people, who most likely live in a very neglected area of a city. Assume that the project has passed the economic, social and environment feasibility tests, i.e., the new undertaking is economically feasible because it offers a very reasonable and realistic plan for people to pay back the loans for their houses, and according to their income. From the social point of view, it will provide decent dwellings with such "luxuries" as electricity, water and sanitation.

From the environmental standpoint, let us say that the new development will be built on an old and consolidated landfill, with proper guaranties for gas venting and burning, leaching treatment, etc. Without a doubt the environment will benefit because a garbage site will be converted in an urban area, with parks, trees, and other amenities and will eliminate a city eyesore.

Water will be provided from a new battery of wells and the new settlement will be connected to the sewage network and then treated. The water wells will extract water from an aquifer which is at that time also being used --- through other wells --- to complement the water supply to the city, whose main source is a nearby river.

So, everything looks fine except for a "small" detail: The average rate of water extraction for this development is calculated as 0.01 m^3/sec, so we are talking about 864 m^3 of water per day. The trouble is, that the average rate of replenishment of the aquifer, from rain and melted snow, is less, as an average, than that quantity. What will be the result in the near future?. The depletion of the aquifer. In other words the rate of extraction of water when compared with the rate of replenishment is not sustainable. Alternatively, a sustainable project would extract water at a rate less than the recharging rate. So, it can be seen that there is a confrontation between economic development and sustainability, which is also to be considered in an EIA.

1.11 The carrying capacity of the environment

We can define this concept as the threshold of stress for the environment that can support population and ecosystems in a sustainable manner. It is necessary to add that the carrying capacity of a region can be changed for better or for worse. Usually it is related with population growth and settlement, for instance when people burn trees to make room for their dwellings. A very clear and unfortunate example are the shrinking forests of Brazil, Indonesia and the Philippines.

1.12 Social, economic and environmental interaction in sustainable development

An EIA must include social, economic, and environmental aspects, but it should be clear that the "environmental" word has to be taken in its broadest context. The main problem, as easily can be seen, is how to coordinate, how to blend all the aspects contained in these three components. This is not an easy task, and besides, there is the problem of how to satisfy goals or objectives that do not have a market value such as beautiful scenery that could be ruined for a project. There is besides, another problem, that we can often have different objectives and some of them contend for the scarce resources.

It is necessary to remember that if we represent each component by a circle, then the only simultaneous solution to our problem is the common space where they intersect, which vertices are a, b and c in Figure 1.1. There is no doubt then that a system has to be found contemplating simultaneously these three aspects and considering all their interactions.

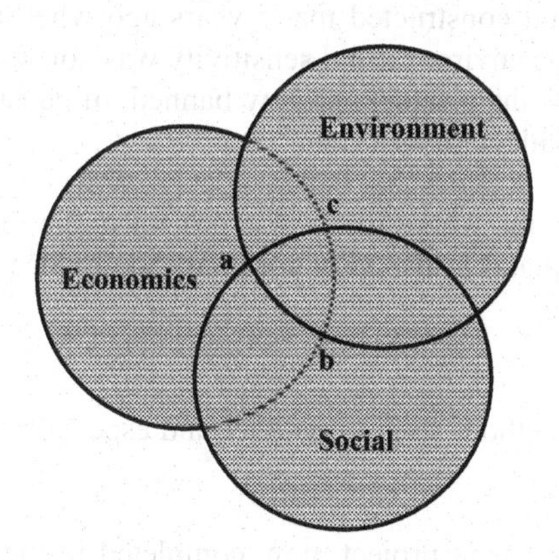

Figure 1.1. **Intersection of three sustainable areas**

1.13 The EIA problem

The EIA tries to find a *balance* between the benefits a project brings and the damages it will produce, in other words tries to compromise between economic development and the damage to the environment.

When only one project is considered, even if it is complex, the problem could be solved without difficulty. But what happens when we have several options or alternatives for the same project --- as is common practice --- and then we have to select only one of them?

Remember, it is necessary to analyze for each alternative the effects it produces, which can be positive, negative, direct, indirect, cumulative, able to be mitigated, etc., and then compare these effects for all the alternatives. Besides, it is necessary to take into account the need to analyze each project from the social, economic and environmental point of view, which is the classical approach, but also considering that the sustainability issue has to be included, which is not so classical.

For most people the word "project" means the construction of something, perhaps a factory, a highway , or a hospital. However there are cases in which the word is just the opposite, that is a project is aimed to destroy something, albeit it would perhaps be fairer to say "to remediate", a word which is extensively used in an EIA.

Most projects of remediation tend to eliminate something that is hazardous to human health and the ecosystem. The great majority are installations conceived and constructed many years ago when the nature of certain actions or when the environmental sensitivity was not so developed as it is today. An example is the production, now banned, of construction parts and insulation material made with asbestos.

There are then situations where the whole project is the dismantling of old installations, and for that it is also very necessary to produce an EIA.

Why?

Mainly because the methods used to recover and especially to dispose of hazardous materials.

There was a very important project, now completed in the USA, under the Direction of the Office of Environmental Restoration and Waste Management. The site comprised a chemical plant and raffinate pits, producing explosives and chemicals such as dinitrotoluene (DNT) and trinitrotoluene (TNT), during Word War II.

Information that follows has been adapted from "Community Relations Department of WSSRAP, a U.S. Department of Energy Project in St. Charles County, Missouri". http://www.em.doe.gov/wssrap/. There are also personal comments from this author dealing with working time at this plant.

After WWII, since 1958 and 1966, the plant was transformed for the conversion of processed uranium ore concentrates to pure uranium trioxide. The Department of Energy mission was to eliminate potential hazards to the public and the environment, and to make surplus real estate property available for other uses. So, why was an EIA needed?

Because there were many potential impacts as follows:

- Many of the buildings had sidings made from asbestos, a known carcinogenic, most of them in a very poor condition, so asbestos could be easily spread into the air.
- There was a quarry, where TNT-contaminated rubble was dumped, and also for the disposal of uranium and radium contaminated building rubble and soils from the demolition of a uranium processing facility in near St. Louis City. This dangerous dumping caused groundwater contamination which was spreading in the direction of the St. Charles Well Field which supplied water to the area.

The remediation project was more than just demolition since it faced a series of challenges with potential impacts, as follows:

- The above mentioned potential spread of asbestos into the air and falling into the ground, with consequences for workers and employees.

- The existence of a school very close to the site posed the danger of asbestos spread into the school population.

- The dewatering of the raffinate pits and the quarry filled with radioactive waste water created serious problems for the workers.

- The disposal of this treated water to the Missouri River also constituted a risk and was bitterly fought by the people living nearby.

- Removal of approximately 95,000 cubic yards of contaminated materials from structural debris, unconfined wastes, contaminated soils, sludge and other solid materials, also posed a risk for air dispersion.

Also, as it is a tornado zone, studies were conducted to assess the influence of this meteorological phenomenon in the area considering the demolition works in the site.

For this reason, before and during the remediation operation many studies were executed to determine impacts and risks, for instance about:

1. Radon control criteria;
2. Radon barrier evaluation;
3. Tests and laboratory analysis;
4. Reprocessing of sludge;
5. Leachate characterization;
6. Radon move in leachate.

The work was successfully completed and the place converted to a recreational area.

1.14 The decision-making process

How is a decision reached for selecting a project, a policy or a program? Once the Environmental Assessment Report (see section 1.15) is finished and submitted to the decision-makers they should have all the elements to proceed with their judgment.

Sometimes the report is reviewed by an independent party that can make suggestions, recommendations, pointing out the aspects that are incomplete, or should be reinforced, or that are missing. Sometimes this independent panel may suggest the rejection of the report because it is incomplete or had insufficient well documented facts and findings.

The decision-makers can ask for more information or for a more complete report according to the suggestions of the reviewer. If it seems adequate, then a decision can be taken regarding the selection of the best alternative, after which it would be desirable to have feedback from the people to be affected by the project; because of this, of course, the decision could be reversed.

EIA has been criticized because it is argued that the decision is very often in favor of the developer because it is he/she that normally pays for this study. For that reason it is strongly suggested to utilize an independent panel of experts who will analyze the EIA and make a corresponding and unbiased analysis and appropriate recommendations.

1.15 The EIA report

A report is usually produced with all the facts concerning the project under study. From this point of view this report is a comprehensive document which must include:

- *Background information* about the project, its main characteristics, geographic location, Terms of Reference (TOR), population and ecological baselines, explaining how this information was obtained, as well as cost of each alternative. Photographs, statistics, etc, to document the statements must be added. This part of the report should also illustrate conditions, alternatives or options, explaining in detail the whole concept and purpose of the project, as well as giving the particulars regarding on what grounds the different alternatives are being proposed. By the same token, if a weight is assigned to each alternative, an explanation should be provided concerning the basis used to assign this priorization.

- *Identification and assessment of impacts.* Here information must be supplied concerning the procedure to identify impacts, and their assessments and significance. In circumstances where the impacts are spatially related, information has to be provided about the way they were determined. For instance, for particulate released from a stack, an explanation and results of the mathematical or other type of modeling used for pollution dispersion has to be detailed.

  Drawings should be provided in order to show, superimposed, the area affected by a particular kind of impact (receptor), and the area covered by the dispersion model. In such cases where impacts involve a supply chain or a path analysis, (see section 5.6), information is needed about techniques and assumptions made. The same applies to temporal impacts. Details should be given about identification of proposed mitigation measures and the degree of its restoration effects, as well as residual impacts.

Many EIA reports are rejected because some of the most important impacts have been omitted or because very weak support is given to justify the adopted values. Remember that impacts need to have some sort of quantitative measure in magnitude and importance, so if particulate from the stack is considered, some cardinal values in a unit such as mg/Nm^3 should be indicated as a threshold.

- *Identification and assessment of mitigation measures* to be adopted, expected values of indicators after or during the execution of this mitigation procedure, and period of time it will last. Information should be given here about the procedure, that is: size and frequency of the sample, frequency of mitigation reports, instruments to be used for measurements, geographical area to be covered, etc. It could be interesting and illustrative to consider mitigation measures taken in similar projects and their results after a period of time.

- *Risk analysis* and explanation of how the main and dispersion values were obtained. *Foreseeable risks* should also be detailed here.

- Externalities, that is actions produced by the project which do not have a market value but that affect people and the environment. See example in section 5.3.

- Compensations to be paid, if necessary, and grounds for their calculations.

- Information provided by public consultation, with a sample of the form used, size of the sample, results from surveys, etc. The method used to collect and input feedback information should be explained.

- Methodology employed for the selection of alternatives or options.

Regarding criteria, information must be given on:

- Procedure to select criteria and reasons for each one;
- Method to determine criteria significance;
- Details of how criteria are grouped, that is, environmental, social, economic, technical, etc.;
- For each type of criterion its main characteristics should be described. For instance in a road project it is necessary in the technical sub-criteria to determine: type of terrain, terrain slopes, geological stability, rivers to be crossed, arable land, etc.;

- If a weight is assigned to each criterion, it is necessary to explain how it was obtained;
- Same for determination of criteria thresholds;
- Care should be taken to avoid criteria redundancy or overlapping and explanation given about methodology and concepts used to demonstrate it.

Internet references for Chapter 1

AREA: EIA
Title: *Methods for interpretation of impacts*
Gives very comprehensive information on:

- Methods for prediction of effects;
- Methods for interpretation of impacts;
- Methods for communication;
- Methods for assessing impacts.;

http://www.adb.org/Documents/Environment/Ino/ino-samarinda-airport.pdf

AREA: EIA
National Environment and Planning Agency - Monitoring Reports - *Dredging and reclamation programme in Kingston Harbour*
Paper produced by National Environmental Policy Act (NEPA), USA..
Details 12 monitoring reports on Kingston Harbor.
http://www.nrca.org/publications/coastal/monitoring_rep/monitoringreps.htm

AREA EIA
Proceedings from the Center for the Study of Environmental Change/Green Alliance Practitioners' Seminar.
Title: *Making Environmental Decisions: Cost-Benefit Analysis, Contingent Valuation and Alternatives*
http://www.sussex.ac.uk/Units/gec/pubs/reps/gralconf.htm

AREA: EIA
National Risk Management Research U.S. Environmental Protection Agency
Title: *Sustainable Technology Systems Analysis - ETO Concept:* Analyzes engineering trade-offs (ETO), as well as economic and environmental trade-offs.

As the paper states:

"While ETO is not proposed as a replacement for life cycle tools and methodologies, it is being developed as a useful methodology for capturing trade-offs in decision-making with the intention of improving the environmental character of decisions without burdening the processing with expensive analytical methods. It is directed to those individuals having to make public and corporate decisions who may not possess the staff or resources for more sophisticated life cycle-based analyses. It is being developed fundamentally as a method to compare two or more products, processes or activities, as opposed to design approaches to improve and existing activity. Currently, NRMRL has developed the concept for ETO and is in the midst of developing a practical methodology". http://epa.gov/ORD/NRMRL/std/sab/ETO_CONCEPT.htm

AREA: EIA
Title: *Guidance on EIA - Screening*
European Commission
This is an extensive document (32 pages) and provides comprehensive information on the process of screening.
Incorporates:

* Screening requirements;
* Implementation of screening in the EU;
* Practical guidance on screening;
* The steps in screening.

http://europa.eu.int/comm/environment/eia/eia-guidelines/g-screening-full-text.pdf

AREA: EIA
Title: *Guidance on EIA - Scoping*
European Commission
This is an extensive document (38 pages) and provides extensive information on the process of scoping

* Scoping requirements;
* Implementation of scoping in the EU;
* Practical guidance on scoping;
* The steps in scoping.

http://europa.eu.int/comm/environment/eia/eia-guidelines/g-scoping-full-text.pdf

AREA: EIA
Title: *Guidelines for the assessment of indirect and cumulative impacts as well as impact interactions*
European Commission
This is a highly recommended document to consult.
It is an extensive paper (172 pages) and provides extensive information in the process of screening.
Incorporates:

- An excellent summary of methods to assess a difficult subject such as indirect and cumulative impacts and impact interactions;
- Provides many examples and graphics for direct, indirect, and cumulative effects, as well as impact interactions;
- Offers a good introduction for the constitution of an EIA team;
- Analyses the different tools and methods that can be applied to projects, and discusses as well case studies for each one, as follows:
  - Expert opinion;
  - Consultations and questionnaires;
  - Checklists;
  - Spatial analysis: Overlay mapping and Geographic Information Systems (gives a very clear idea of how GIS is used in this context);
  - Network and system analysis;
  - Matrices;
  - Carrying capacity or threshold analysis;
  - Modeling.
- Scoping. Extensive comments about boundaries and especially considering past, present and future impacts of the project;
- Information needed to assess the impacts;
- Impact interactions;
- Mitigation measures;
- Transboundary impacts;
- Magnitude and significance of impacts;
- Monitoring;
- Presents ten case studies.

http://europa.eu.int/comm/environment/eia/eia-studies-and reports/guidel.pdf

AREA: EIA
Title*: Module 10- Environmental Impact Assessment*
Published by the Forum for Economics and Environment.
Presents a very good table which can be used as a guide for determining attributes for projects related to air, water, land, ecology, sound, human aspects, economics and resources. Another interesting feature is the

classification of EIA's by The World Bank category regarding the need to perform EIA's for different types of projects.
It also shows in a diagram the World Bank EIA process.
http://www.econ4env.co.za/training/MODULE10a.pdf

AREA: EIA
Title: *Environmental impact assessment of irrigation and drainage projects*
Authors: T.C. Dougherty & A.W. Hall
A very comprehensive report (94 pages) on EIA, prepared by FAO.
http://www.fao.org/docrep/V8350E/v8350e00.htm

CONCEPT: EIA
Title: *Environmental Impact Assessment for the natural gas distribution network in Lima and Callao.* Transportadora de Gas del Perú S.A.
http://www.camisea.com.pe/EIAPdfIng/CDEnglish/Vol%20I/PART_I Text.pdf

CONCEPT: Environmental management
Title: *EMS as an Opportunity for Engaging China's Economic Development Zones: The Case of Dalian*
This paper illustrates the value of environmental management systems China's economic development zones by using the case of Dalian
http://www.uneptie.org/pc/ind-estates/casestudies/Dalian.htm

CONCEPT: Methodologies
Title: *Environmental impact assessment methodologies description and analysis and first approach to environmental impact assessment methodologies application*
oms.org/muwww/fulltext/repind51/environ/environ.html

CONCEPT: EIA Report
Title: *Quality analysis of the EIA draft report for the Baku-Tbilisi- Ceyhas oil pipeline: Turkish part*
http://www.bankwatch.org/downloads/btc-esia-tur-analysis.pdf

CHAPTER 2 - PROJECTS AND IMPACTS

There is a rational process for alternatives or projects selection that applies to the different methods employed. This process involves the following steps:

2.1 Defining the objectives of the study

It could be that the project calls for the construction of a petrochemical plant, and in this circumstance clearly the objective is to obtain some monetary benefit. Sometimes it is not possible to place a commercial value on the objective; this could be the case of planning the construction of a park for recreation purposes. In this case the objective is to obtain an intangible benefit which is the construction of an amenity. On the other hand the objectives, whatever they might be, need a *time framework*, or in other words they have to be *time-dependent.*

In many cases the objective plays a fundamental role. For instance in the construction of a petrochemical plant probably some sort of evaluation techniques will be used considering the economic aspects of the project, (commercial objective), such as a Cost-Benefit analysis, a Net Present Value calculation or the Rate of Return. In other circumstances there is not a monetary value to be assigned to the objective as in the case of the construction of a park, (social objective), in this case, probably a tool called Cost-Effectiveness is the method of choice.

2.1.1 Projects, alternatives or options

Very often it is possible to select amongst projects or alternatives that stand alone, but sometimes it is necessary to consider that projects can be linked with some sort of association that precludes the execution of one of them without the execution of another.

It could also happen that just the opposite occurs because of the existence of exclusive projects. Besides, there are some elements that have to be considered in real life situations, such as:

- Many projects take more than a year to be completed and cash disbursement for their execution is done at regular intervals or with a certain rhythm. This sort of cash flow problem introduces some difficulties in comparing projects since not all of them have the same cost, let alone the same funds schedule.

- Then, it also comes to different durations. To compare projects from the economical point of view one has to take into account the year the costs and benefits occur, and then calculate the corresponding discount at the present time, as well as the especial characteristics of each project regarding such things as payback periods.

  There is no need to expose here the very well- known financial techniques but just mention the kind of techniques that can be utilized.

  For instance use can be considered of the Cost-Benefit Analysis (CBA) if the value assigned to the target is known, but if this is unknown, such as in the case of a social project involving the construction of public hospitals for instance, obviously there is no a value for this objective.

  In these circumstances it is possible to employ Cost-Effectiveness Analysis (see section5.4).

  If the project or projects involve special issues it is quite possible that market rates cannot be used to compute the financial analysis, because usually these rates are distorted for many different reasons. In these circumstances one should use social values that can be derived from shadow prices (see Appendix, section A.7).

- Another real life factor is related with the time, speed and sequence that is inherent to each project. It is extremely unlikely that two competing projects will develop the same construction schedule that materializes in the form of the investment curve of each one (the "S" curve).

  For this it is necessary to take into account the bar chart or Gantt diagram for each project, and from there deduct the different degrees of completion for each main activity and for each period of time.

  Recognizing that in real life situations it is necessary to coordinate funds available for construction in a certain period of time with the construction schedule and its percentages for each main activity, one can understand the difficulty of the problem, and remember that this is done for each competing project.

- Last but not least, data must be obtainable considering the availability of certain resources, which normally are accessible as a pool for all projects, such as City Hall personnel of engineers, draftsmen, inspectors, etc.

2.1.2 Why are there alternatives to a single project?

Up to now no mention has been made of a very important fact: cost.

In EIA it is not often that there is only one project with no alternatives to analyze. This is so because even though with a single project, it usually has different alternatives or options. Normally each alternative has a different cost, even with the same project, because every one will have a particular route, will produce more or less pollution, will involve different works, will traverse different populated areas, etc. As an example, a highway project between A and B can have say four different alternatives, according to the following aspects:

Alternative a): The shortest route, but it needs the construction of a tunnel 3 km long, and passes through an area of cultural heritage for a native tribe;

Alternative b): A little larger than a), does not have a tunnel, but needs the construction of a bridge in an area subject to periodic flooding;

Alternative c): The longest route by many kilometers, and also the least costly, but it passes through a national park and in an area with high slopes;

Alternative d): A little shorter than c) but traverses a geological faulty region prone to landslides and tremors.

Each one of these alternatives will do the job, but at a different cost, involving different environmental issues, facing various social problems and with different risks. Our problem is: How do we evaluate each alternative, compare with the others, and select the best?

2.1.3 Inventory of alternatives or options

To achieve the objective a plan, a project, is needed. More often than not, even a single project can be broken down into different alternatives mainly considering environmental and social factors. As an example, it could be that the cheapest route for a certain project is to build it through a wetland or a populated area. Although the most convenient, economical and shortest way, probably is unacceptable because of the disruption it will cause in the natural area, problems or even risks to the population, etc, so other routes or alternatives are devised, most likely at a higher cost, but with less impact to the environment and to people. These comments are aimed to indicate that there could be several alternatives or options to reach the objective/s.

Even in the case of the construction of a park it could pose ecological problems, such as the perturbation of wildlife or alteration of the environment because of the presence of too many visitors, which is normally the case in some natural areas, where the presence of numerous people, walking off the specified trails, with the inevitable garbage they produce, can damage the area. Therefore in many cases it is necessary to have a list of alternatives or options to achieve the proposed objective.

In some cases and for whatever reasons some alternatives are preferred over others, so a weight is assigned to each alternative to indicate this preference, but this is not an assurance that the alternative with the highest score will be selected, since it depends of its degree of compliance with the established criteria used for selecting alternatives. This preferred weight is called *a-priori weighting.*

2.1.4 Identification of criteria to compare alternatives

Criteria or attributes are components or considerations that can be used to compare alternatives. Values are determined for each alternative and for the whole set of criteria, one value per criterion, in a process called *scoring.*

Then, a score, corresponding to an alternative and a criterion will reflect, in a certain scale, how well the alternative meets the objective. The complete set of scores, involving all alternatives and all criteria, is called a "performance matrix". In the interaction between an alternative and a criterion, there is then a coefficient that evaluates the alternative with respect to the criterion. To obtain these scores different techniques can be used, but most of them rely on expert judgment or expert opinion. Different methods have been developed to get this expert opinion in an unbiased manner, since the chance of individual preferences is always present. A technique called "Delphi" (see section 4.3.7) has proved to be very useful in solving this problem.

If this performance matrix alternatives are in columns, and criteria depicted in rows, then, for a certain alternative or option, its corresponding column intersects all the rows, and will indicate how well this particular alternative meets the objective when all criteria are considered. Conversely, a row representing a criterion will indicate how well the objective is met for that criterion, when all alternatives are considered together.

2.1.5 Projects' characteristics

There are without doubt a great variety of projects than can be subject to EIA and they can range from the construction of a pipeline, to the installation of a thematic and amusement park or the development of a natural park.

However, most of these undertakings have a common denominator, and it is that they will *alter the environment* in the area where they will be constructed. For that reason, in the "scoping" stage of the EIA it is imperative to analyze and develop a set of alternatives that can decrease the effect on the environment or to the human health.

2.1.5.1 Selection of alternatives

Frequently, in simple scenarios, the performance matrix alone is used to make a selection of the most beneficial alternative. In general, the higher the score the better, therefore, if one column, i.e., a certain alternative, shows higher values in its column than any other option, most probably this is the best choice.

In general, this does not happen, since an alternative can have the highest values with respect to the other alternatives for one criterion (the one with the largest value in the criterion row), but also have lower values for other criteria. One can say, well, a high value for one criterion compensates with a low value in another. Yes, it could be if all criteria had the same weight, but this is not generally the case, therefore it is not often that this type of "compensation" occurs.

2.1.5.2 Public opinion

Normally the decision about the selection of one alternative or other is made by the decision-makers and stakeholders. These people act with the information provided by the performance matrix and with the results derived from applying certain techniques. However, it is a good practice to get feedback from the population who will be affected by the project. Most certainly they can contribute with information which was not considered by the stakeholders.

There is a movement nowadays to consider very seriously this feedback from public opinion, coming from meetings, surveys and polls. In reality, this type of consultation has to be included as a criterion, not incorporated as feedback after a decision has been made. In this way what the people want and feel is used to gauge, together with the other criteria, how the alternatives meet the objective.

2.1.5.3 What are the problems for appraisal?

In the old days, before the awakening of the necessity of EIA, the selection was simpler. If we had to build a rail passenger route between two points, we laid the tracks as straight as possible, studying of course the different alternatives regarding intermediate population locations close to the line, studying the geology of the route, and naturally the different costs.

One example is the laying of the tracks for the railway between Salt Lake City and San Francisco in USA. The last spike was nailed on May 10th 1869 on a route north of the Salt Lake. This was a path that had some geological problems and was not the shortest, for the most direct way would have been to lay the tracks through the Salt Lake, but at that moment this option posed some technical difficulties.

As a matter of fact some years latter, the new, and present day route, cut through the lake. At that time nobody cared about the ecological effects that such an undertaking would have in the lake ecology. Most probably, nowadays that project through the lake would have been rejected, because some parameters other than the economy of the line, are also to be considered.

However, even today, in many projects, some other aspects such as public opinion or sustainability are seldom taken into account. From this point of view people's perceptions about noise to be generated in a new highway, or the uneasiness produced because the highway will separate two neighboring areas, or the aesthetic effects of the new undertaking, have to be carefully considered and studied. In other words, people's opinions are as important as the more classical issues, and must contribute to the appraisal.

All of this introduction leads to the fact that it is necessary to determine what are the facts, or criteria, or components that will be used to quantify the different alternatives. This step is the first one in a project evaluation, and is called *scoping*. In scoping one tries to identify the key aspects of the project and establishes its terms of reference. It is also here where the decision is taken about performing or not an EIA. This decision in turn, depends on the type and size of the project.

However, whatever the criteria, they involve comparing components that belong to different fields, with different units of measure --- and in many cases without any --- such as the above mentioned esthetic pleasure of bird watching, so one can appreciate the difficulty that this comparison poses. Besides, not all criteria have the same weight, for it is customary to evaluate one criterion against another and establish their relative importance. This topic will also be analyzed in section 5.8.

2.1.5.4 The baseline

It is essential to have a good inventory of the situation before the commencement of the project, in order to be able to make future comparisons between conditions after the changes and effects created by a project, and the original ones. This is called the *baseline*, and it applies not only to the environment but to the population as well.

As a matter of fact these changes will be measured considering the difference, in many diverse aspects, between these baseline conditions and the situation in existence after the implementation of the project. Maps, photographs, data, opinions, etc will help in establishing this baseline, and the use of GIS will be very useful for this purpose. See section 5.11.

The baseline is not constrained to learning *what* it is but also *how* it is determined. In this context an effort has to be made to understand how the ecosystem works in a particular area, how it affects other areas, and how it relates with the human population. It is important to gather information about the past history of the area to be affected by the project such as heritage sites, the existence of sacred grounds and sites according to religious or traditional beliefs, the history of periodical natural effects such as hurricanes, floods, earthquakes and tremors, etc.

2.2 Impacts caused by projects

2.2.1 Impact definition

An impact can be defined as the change of some conditions in human health and/or in the ecosystem caused by the development and implementation of a project. This is clear, but how is the importance of an impact measured?

An impact has magnitude, i.e., it could be large or small, and can also be important or not. As a consequence the significance of an impact is the result of these two aspects, i.e., magnitude and importance. An earthquake with a magnitude 7 on the Richter scale is serious, but if it happens in a desert area, it does not produce any damage, so its material importance is reduced A low magnitude earthquake in a heavily populated area can be extremely important because of the damages it can cause.

2.2.2 Effects caused by projects

There are a great variety of effects that can be caused by the proposed project or projects. They involve environmental, economic, social and sustainable aspects. A classification of impacts follows:

2.2.2.1 Positive or adverse

Positive effects are related with a *benefit to the people* or to the environment. Adverse effects relate with impacts that can *deteriorate* the way of life of people involved or that cause damage to the environment. Some projects generate positive effects but at the same time create new problems or adverse effects.

Many projects, like the construction of a gas pipeline between two points, might involve intruding into environmentally sensitive areas, for example bird sanctuaries, national parks, forests, wetlands, etc. This type of project produces benefits not only because it will allow a country to export its gas in this very convenient and economic way, but also by providing employment and a better living to people along the route.

But there also can be secondary positive and negative effects. The possibility that the fluid could be used for heating in domestic furnaces, or in industrial plants, instead of burning oil or coal, will greatly reduce the air pollution in a city. This has happened for instance in the City of Santiago, Chile, which now receives gas from a neighboring country through a gas pipeline.

It also can disturb natural life or destroy hectares of forest or alter the landscape in a mountainous area because of the construction of a service road, boosting stations, and so on.

2.2.2.2 Primary, secondary, tertiary.

Direct or primary effects are the immediate consequence of the project. Indirect or secondary effects are those produced by the direct effects. Many projects --- hydroelectric plants for instance --- create several positive direct effects such as encouraging economic growth by production of cheap energy for industry, and the improvement of living conditions due to the fact that electricity can now reach more households. Because of the dam in a reservoir a river discharge can be regulated stopping devastating periodic flooding, and encouraging tourism in the lake behind the dam. The water can also be used for drinking purposes as well as for irrigation, increasing the arable land.

However the same project can also generate several adverse direct effects. An example is the Aswan Dam on the Nile River, because there is no longer an annual flood. As a result, the fields along the river are deprived of the very rich mud carried from far away areas, and as a consequence fertilizers need to be used. This not only creates a cost --- inexistent before --- but also can have adverse effects because of the introduction of nitrogen into the soil. It also produced the loss of many priceless structures --- although fortunately many were salvaged --- when the Nasser Lake created behind the dam, covered them.

There are also secondary adverse effects such as the disappearance of certain fish species in the Nile Estuary because the water no longer carries the rich food found in the mud for these species. In turn, there exists a third effect, for the natural habitat for birds has been altered because of the absence of fishes on which they can feed.

There are some projects that *apparently* could generate only benefits without any major drawback. For instance there is a solar-hydropower blueprint in Egypt for extracting water from the Mediterranean Sea and piping it to the Qattara depression, in Central West Egypt, taking advantage of the level difference, to generate 1,800 mW. (There is a similar project in Israel discharging into the Dead Sea). Apparently this project can only benefit the region, since it will bring electricity to a remote area, however it also creates some effects which could be adverse, as for instance:

- The Qattara depression is home of a large aquifer, and it is unknown how the salt lake to be created in the Depression will alter this aquifer and the existing oasis, which are vital for the survival of many people in this area;
- As any other large reservoir, the artificial lake could induce seismic effects;
- It is also unknown how the lake will perturb wildlife in the area;
- There are some archeological sites in the area which could be lost because of the lake;
- Most certainly the intake built in the Mediterranean will produce sand drift and a high turbulence in the water. This construction might create secondary effects which could influence the fish habitat.

2.2.2.3 Measurable or indeterminate.

Some effects are measurable while others are not. When considering measurable effects --- such as electricity generation --- a value can be obtained, but sometimes this is not possible even if there is one. For instance the correlation between large masses of water behind a dam and the possible earthquakes they might trigger is not known, let alone how to put a value to its intensity. Examples of non-measurable effects are for instance the pleasure of bird watching, or the relaxation of sailing in a river, or the emotional stress produced in people when their town is swallowed by a new reservoir.

2.2.2.4 Apparent impacts

Sometimes the benefits are only apparent. In Los Angeles, highways permit the fast movement of people and merchandise. They were built to avoid delays and jams, however --- and this is a known fact --- highways feed their own growth, since more and more vehicles use them everyday because of their convenience. This, sooner or later produces traffic jams, which calls for the necessity of building an additional lane or another highway, and so on, unless some measures are taken.

2.2.2.5 *Cumulative*

There could also be cumulative effects of a project in a certain area. For instance something as innocent a project as the construction of a golf course, can generate small environmental effects, such as the desiccation of a lagoon, the creation of artificial dunes, the logging of some trees, the perturbation of a bird's habitat, etc. But when these effects are considered as a whole , the total impact could be significant. The appraisal of cumulative effects is not easy, because of the difficulty in delimiting or defining the geographical area where the effect takes place, that is, its *spatial boundary*. A classical example of spatial boundary is the area contaminated by air emission from an industrial plant.

There is also a *temporal boundary*, since different effects can have different durations. From this point of view it is important to examine past, present and future consequences. In section 5.6 this cumulative issue regarding past, present and future actions is analyzed.

2.2.2.6 *Able to be mitigated or not in a greater or lesser degree*

After its execution some projects can cause worsening in the environment, for example disturbed land, construction wastes, open trenches, or cause a deterioration of the quality of life of people living nearby. An example could be the construction of a metropolitan rail transportation system (Light Rapid Transit or LRT). Train traffic can create severe discomfort and distress in houses near the tracks because of the high level of noise and vibration. Of course, some remediation or restoration work can be done, such as the construction of sound barriers, but usually there is always a *residual impact* that cannot be eliminated.

In other cases it is quite impossible to remediate a situation, such as for instance the big excavation left after the exploitation of an open pit mine has finished.

2.2.2.7 *Residual impact*

Once all the *mitigation measures* have been applied to reduce an impact there could be a residual impact. It is necessary not only to identify this residual impact but also to evaluate it in magnitude, importance and duration, as well as measures taken to monitor its trend.

2.2.2.8 Spatially related

The spatial extension of an effect relates to its *diffusion*. For instance the Chernobyl disaster caused radioactivity to an extent hundreds of kilometers from the site of the explosion. Projects such as the construction of big waste incinerators can contaminate the atmosphere for many kilometers from the plant. Naturally, the way contamination spreads largely depends on atmospheric factors such as winds, mountains, low or high pressure areas, etc.

There are mathematical diffusion models that can predict the *extent* of the contamination and determine where the particles would settle, but in many cases a *health risk is also involved.* For instance, mining projects and refining plants close to cities could be endangering the neighborhood not only because of air emissions, but also considering that some chemicals in the tailing ponds can reach the underground aquifers through gashes and holes in the plastic protection and contaminate them. If the city is extracting water from these aquifers, there could be very serious health consequences.

Regarding spatially related effects there are also *transboundary effects.* These do not always mean environmental or health damage, for they can also have economic consequences. As an example, the construction of a dam most probably will affect part of a river basin, or perhaps the whole basin, involving more than a country. A hydro-project downstream in a river can affect the maximum output of another hydro-plant built upstream, separated by hundred of kilometers. This can be actually seen in two dams built in South America on the Paraná River. The upstream dam is the Brazilian hydro-station of Itaipú --- the largest in the world --- while the downstream dam belongs to the Argentine hydro-station of Yaciretá. Both undertakings have to be related since an increase in the level of the Argentine reservoir could affect the head of the Brazilian dam, and therefore its electrical output.

Another example of these transboundary effects is the production of acid rain in Canada, generated by industries in the United States.

2.2.2.9 Temporal related

There are effects that are *time-dependent.* This is the case of emissions produced by automobiles. Over time, and depending on atmospheric variables, the effect can disappear. Some projects have to also consider for how long their effects will hold. Extensive logging could be an example. For instance how much soil erosion will take place when an area is logged, and until a new planted forest decreases the erosion rate?

2.2.2.10 Irreversible or reversible

Mining operations produce irreversible impacts, since we can not restore the baseline conditions in a near or even a distant future. Some huge projects such as the construction of the pyramids in Egypt can also be considered to have produced an irreversible visual impact in the desert, and although these constructions can always be demolished to restore the visual landscape, it is very unlikely that it will happen, since then we would be altering the present day baseline by destroying a man-made cultural heritage.

The loss of forests in many parts of the world to make room for agriculture, housing and roads, is practically irreversible, for if it is true that theoretically they can be restored, in reality it is not going to happen in many, if any, cases. Reversible impacts are those that naturally or by human intervention can be eliminated after a certain period of time. For instance if bird population in a forest diminishes because of the noise caused by a project construction, most probably this population will return in the future if the origin of the noise disappears. Another example is the cleaning of many rivers that were completely polluted and practically without fishes. After remediation works take place, there will probably be generated a fish population whose size could be close to the original one.

2.2.2.11 Likelihood of the impacts

In many cases one is not certain that an impact will occur. For instance what is the probability that the construction of an industrial park will have a negative impact in its neighborhood?

If the project is the construction of a highway in a residential area, for sure many people will complain because the highway will impact their easy access to other parts of the same area, or for whatever other reason. However, it could very well be that after the highway is in operation the people are happy about it because their access to downtown is now much easier and faster, a fact that was not appreciated before.

2.2.2.12 Unexpected impacts

Of course, nobody can predict the effects of something unexpected or whose *eventual existence is ignored.* Unfortunately there are many extremely serious examples of unexpected impacts. For instance the Thalidomide drug had terrible consequences that nobody could predict. The discovery of Freon 12 (a fluid used in domestic refrigerators and in car air conditioning systems), now banned, is blamed, at least in part, for the gradual disappearance of the ozone layer. As another example, if a project calls for the construction of an off-shore oil platform, there are some unknowns such as the effect that a hurricane can have on oil spills in the sea.

Some unknown effects have been however beneficial. Take for instance Aspartame®, a sweetener that replaces sugar but only when added in cold products such as ice cream, soft drinks, etc. Its discovery was fortuitous and completely unexpected. In these cases thorough and systematic studies are needed to investigate the possible consequences of an action. See for instance section 1.13 where the care taken in dismantling an old industrial plant in the USA is discussed. Because there were many unknowns, nearly fifty meticulous studies were conducted on different issues, with some of them breaking the ground with new investigations.

The problem with many impacts is that they could be dangerous *when combined.* For instance take the emission of SO_2 (sulfur dioxide) from many industrial plants and powerhouses. The gas itself when breathed has severe effects in the heart and in lungs, and can produce respiratory illness. When it is combined with water present in the atmosphere it produces sulfuric acid and acid rain. This acid rain is blamed for the acidification of water bodies and the corresponding impacts in the fish population.

The problem with the unexpected is that they could be *time-dependent,* meaning that their existence is not in evidence until some time has passed, (again, the Thalidomide example), and of course they also could be spatially spread.

2.2.2.13 Risk effects

To determine risk, a measure of the impact is usually obtained and it can be stated for instance in a sentence like *"It affects one in 100,000 people".* This is a *measure of risk* and is a *mean value,* and usually a *standard deviation* can be calculated which in essence means for instance that in 95 % of the cases there is a probability that an impact will be within the limits established by the standard deviation.

The standard deviation can be used to determine the *significance* between two different impacts. When their respective mean value and standard deviations are known, one can ascertain which of the two is riskier. When this mean value is imprecise or indefinite, and as a consequence limits cannot be established for the standard deviation, then there is *uncertainty*.

A project can also involve some risks of very different nature. As mentioned, the environmental risk of a leakage can turn an aquifer useless, but risks are not only environmental-wise; as an example consider that there could be a *political risk* when a pipe line is vulnerable to being cut or blown up by warring factions, as happened in August 2003 in the Iraqi pipeline transporting oil to Turkey. There is also a *geological risk.* Many years ago, in 1963, the Vajont Dam in Northern Italy and the biggest in Europe, broke up as a consequence of boulders falling into the reservoir, because of lack of stability.

2.2.2.14 Residual effects

Mitigation involves taking measures to remediate totally or at least in part the damage produced by a project. Many times the consequences of a certain impact stay even after mitigation has taken place. These are called residual effects, and might be able to produce secondary impacts. The determination of the magnitude and importance of these effects is called *significance.* Monitoring is closely related with residual effects since one of its main features is to keep a close eye on them and make sure that they stay within the stated limits.

2.2.2.15 Population impact

In a preparation of a large project it is necessary to consider the impact that workers and staff will have in the society were they will be immersed. Normally there are two kind of situations:

- The construction site is in a *desolate area* in which provisions have to be taken for the accommodation of the personnel, generally in trailers. Of course, there is a need to build a sewage network as well as a sewage treatment plant. This appears to be quite excessive, however, considering that in a large project there could be a construction camp with more than 800 people, this measure makes sense.

- The construction site is *near a town* which can absorb the influx of workers and staff without many problems. But in some occasions there is a little town or a village whose population is perhaps a little more than the camp population. In this circumstance the problem could be very acute because even if there is camp accommodation for workers, senior personnel and staff usually lodge in town.

This impact could be important especially because of children from the senior and staff personnel attending school. It can generate problems of room *overcrowding* which will most probably have a negative impact in the local population, as well as an increase in prices that will have the same effect. Another aspect to be considered as an impact, and as a negative one, is the *potential for crime* in town if it is situated near the camp. This is a very common problem because of the existence of hundreds of young males, staying in camp for long periods of time, with some access to alcoholic beverages and temptation to go into town.

This author has worked in camps where workers as well as staff were forbidden to go to a nearby town even for visiting, under penalty of immediate dismissal, a measure that was strictly enforced.

2.2.2.16 Interaction between impacts

Sometimes impacts belonging to the same project or to different projects can interact and in so doing *create a third impact.* To illustrate this case, assume that there is a coal fired powerhouse and close by a cooling tower pertaining to a meat packing plant. Both produce direct effects: The powerplant can spew H_2S that generates health problems, while the cooling tower creates a micro clime with very high humidity. When both effects interact, acid rain can be produced.

2.3 Conclusion on impacts

In conclusion we realize that a project can generate *beneficial effects and adverse consequences* in different scenarios, in *various spatial and temporal boundaries*, with a *certain mitigation* and with a *residual effect* after mitigation. As in any other things in life, there are pro's and con's, however it is a fact that whatever the nature of a project *it will disturb the environment,* most of the times on the negative side. A check list of impacts is presented in Table 2.1, that can be used to make an impact inventory for any project. In this example it is assumed that there is air contamination from a certain undertaking.

Table 2.1 **Check list for impacts**

Impact type	Yes	Extent of impact	Degree of mitigation
Positive		There will be generation of employment in an industry with a high multiplier effect	
Adverse	x	Produces air pollution with SO_2, NOx and HS_2	With the new electrostatic filters the contamination will be just below the maximum thresholds, except for NOx. Studies are being conducted to ameliorate this effect and bring it within acceptance limits
Primary	x	Affects human health	
Secondary	x	Produces acid rain	
Tertiary	x	Death of fishes in the river	
Measurable	x	It is very easy to take samples and to measure concentration	
Indeterminate			
Apparent			
Cumulative	x	On top of health hazard it will provoke acid rain, especially SO_2	
Able to be mitigated	x	Certainly. Studies are conducted to determine the best system.	
Residual impact			
Spatially related	x	Diffusion studies show a plume extending 25 km from the plant site with decreasing intensity and concentration	
Temporal related			
Reversible			
Irreversible			
Likelihood	x	There is no doubt of the effects of this emission, although the wind factor can contribute to its dispersion	
Unexpected impacts		Unknown, but from studies in similar plants installed many years ago and without the benefit of the advance filters that this project is using, it could be concluded that no further effects are foreseen	The construction of a higher stack could decrease unexpected impacts effectsbecause there will be an increased dispersion of emissions
Risk effects	x	There is some risk due to the nature of gases released	Filters will be installed in the smokestack
Residual effects			
Population impact			
Interaction between impacts	x	Emissions could interact with another industrial plant releases	

Internet references for Chapter 2

CONCEPT : Impacts
Title: *Appendix C - Environmental Impact Statement - Significance criteria Rural Utilities Service* - U.S. Forest Service
Highly recommended publication

It considers:

- Magnitude of the impact (how much);
- Duration or frequency of the impact (how long or how often);
- Extent of the impact (how far);
- Likelihood of the impact occurring (probability);
- Intensity of the impact;
- Intensifying factors;
- Criteria for rating impacts;
- Significance definitions for different type of impacts as follows:

Receptor	Effect
Soil	Erosion
Soil	Contamination
Farmland	Loss
Air	Quality degradation
Water	Downstream water flow reduction
Water	Surface water flow reduction
Ground water	Water quality reduction
Wetland	Degradation
Floodplain	Damage
Terrestrial resources	Biological resources degradation
Recreation	Degradation
Cultural resource	Degradation
Hazardous waste	
Solid wasteland use	Conflicts
Agricultural land	Loss
Traffic	Congestion
Noise	
Alteration of visual quality	
Human health and safety	
Socioeconomics	
Changes in income	
Changes in tax base	
Residential relocation	
Tax	Assessment changes
Environmental justice	

CONCEPT: Impacts
Title: *Nepal: Impact identification of Langatang Kohla Hydropower project*
A United Nations publication with good definition on impacts as magnitude, extension and duration. Also information on:

- Uncertainty in impact prediction;
- Comparison of alternatives;
- Key elements for assessing impact significance;
- Example on impact identification for a hydropower project in Nepal;
- Example on prediction and determination of significant impacts.

http://www.unescap.org/drpad/vc/orientation/m8_anx_4.htm

CONCEPT: Impact
Title: *Impact prediction comparison of alternatives and determination of significance*
http://www.unescap.org/drpad/vc/orientation/M8_5.htm

EIA REPORT
Title: *Sasol Natural Gas Supply Project*
Analysis of a natural gas project in Africa. The paper makes very good comments about difficulties encountered in the route selection, as well as a detailed description of the procedure followed.
http://w3.sasol.com/natural_gas//Environment/RSA%20Document%20PDF/RSA%20Draft%20EIA/Chapter%203.pdf

EIA REPORT
Title: *Technical Memorandum-Environmental Impact Assessment (EIA) Report*
Hong Kong Environmental Protection Department.
Describes objectives and content of an EIA report. Especially important is the evaluation of the residual environmental impacts.
http://www.epd.gov.hk/eia/legis/memorandum/text4.htm

CHAPTER 3 - CRITERIA

3.1 Criteria, Attributes or Components

Criteria can be considered as *parameters used to evaluate the contribution of a project to meet the required objective.* There is no procedure to determine the number and class of criteria used for the appraisal process. Because of this, very often two or more criteria express the parameter, and for this reason, and assuming that the different criteria are linearly associated, we discuss in section 3.2 a methodology called *Factor Analysis*, to minimize the number of criteria to take into account *without loss of information.*

There are some criteria that are of general use whatever the type of project, but others are more project dependent. In this book a set of criteria are proposed and it is believed that they can be used in a large number of different projects. The way to proceed is to find a relationship --- usually a cardinal number --- between each alternative or project and each criterion. This relationship establishes the *contribution for each alternative,* regarding the corresponding criterion, to meet the objective.

The way the criteria are used largely depends on the system utilized for appraisal. For instance the Analytical Hierarchy Process (see section 5.8) compares each criteria with another and establishes a number between 1 and 9 *expressing a preference.* When alternatives are compared between themselves the same system is applied, and then these figures are multiplied by the weight obtained for each criterion (see section 5.8).

In Mathematical Programming, the relationship between projects and criteria can be expressed for any number in any unit, provided that the same unit is kept for a row, albeit the next row can have something completely different unit-wise. Also, in Mathematical Programming a limit or limits can be established for each criterion These limits are called *thresholds*.

The following listing identifies the criteria that can be employed to compare alternatives.

3.1.1 Technical criteria

These are criteria related to technical characteristics of each alternative or project. Suppose that a project calls for the selection of four different alternatives or a combination involving four different systems to control the discharge of contaminated water into a river. Assume for instance that we are controlling only one criterion, which is related with the content of BOD_5, a measure of contamination (see Glossary).

There could be for instance a number such as 175, specifying that alternative number 2 produces waste water with a contamination of 175 mg/liter. Another alternative, say number 4, will indicate for the same criterion a waste water with a content of 152 mg/liter, and so on. Obviously these numbers means nothing if a *limit or threshold* is not established, since the fact that 152 is smaller than 175 means less contamination, but *it does not necessarily indicate that this lower value is acceptable.*

In this context our threshold could be for instance 180 mg/liter, so, for this criterion, an alternative discharging more than 180 mg/liter does not comply with this restriction, and with this threshold now it can be said that alternative number 4 is better than alternative number 2. Other technical criteria *specify capacities.* For instance in a project utilizing raw water we can establish a maximum amount of water that can be treated, because that is the capacity of the water treatment plant.

3.1.2 Environmental criteria

Here criteria normally relate with thresholds that *are not to be exceeded.* Examples of environmental criteria are:

Water use;
Recirculated phosphorus;
Discharges to the air;
CO_2 discharge;
CO discharge;
NOx discharge (see Glossary);
Particulate;
Discharge of toxics into water;
Hot water discharge;
Fossil energy use;
Number of trees to be cut;
Water bodies crossing;
Number of km. of visible effects (for instance a pipeline over the ground);
Land use;
Nearby or crossing wetlands;
Indigenous forest crossing;
Degraded forest crossing;
Natural park crossing;

Noise production;
Impact on groundwater;
Fauna migration;
Wildlife sanctuary;
Atmospheric emissions;
Hazardous waste;
Wetland crossing;
Marshes to be crossed;
Chemicals to the waste water;
Deserts crossed;
Dry land crossed;
Kilometers of service roads;
Visual pollution;
Erosion;
Biological effects;
BOD_5 content (see Glossary).

3.1.3 Safety criteria

Here we are dealing with criteria whose thresholds are expressed in percentages. As an example it is possible to state that the risk of soil contamination is very low, possible at 2 %, or that the risk of social unrest is relatively high at 15 %. The model in multicriteria analysis (see section 5.9) can examine these cases, and if this one were the only criterion, possibly the model would select the alternative that offers the least risk. But usually this is not the case, since there are more than one criterion, so the *model must find a compromise* between the different criteria such as:

Social unrest;
Energy security;
Security risk;
Sabotage risk;
Geological risk;
Political risk;
Risk of soil contamination;
Seismic effects;
Etc.

3.1.4 Social criteria

Percentages can also be used for social criteria expressing *people perception on different issues*. For instance a project can receive a public acceptance of 45 per cent while another project gets a 70 per cent approval rating. Examples of social criteria are:

Citizens' evaluation and opinion for and against the project;
Public acceptance;
People affected by the project;
Persons to be resettled;
Populated area at less than a certain distance from the project's site;
Public health and safety;
Heritage conservation;
Crime during construction.

3.1.5 Economic criteria

These are criteria that *express economic facts*, for instance the hectares of ploughed fields, the hectares of agricultural land impacted by the alternative, or the traffic flow expressed in number of vehicles per hour. Examples are:

Arable land;
Corn fields;
Direct economic benefits;
Indirect economic benefits;
Traffic flow;
Traffic volume;
Urban traffic;
Commercial forests;
Economic efficiency;
Market competitiveness;
Production unit cost;
Resource use efficiency.

3.1.6 Construction criteria

Again, these are technical criteria, such as the slope of the terrain or the distance to a road. Examples are:

Geological faults;
Geological stability (or lack of it);

Difficulty for access;
Lack of water;
Type of soil;
Engineering difficulties;
Logistics;
Open cuts;
Transmission distance.

3.1.7 Spatial criteria

These are related with the *spatial effects* of the alternative or project. If the project extends over some length, the effect that the distance has on it can be expressed here. Examples are:

Transboundary effects, that is to say effects across the borders between nations;
Existent right of way (servitude);
Diffusion of the concentration of a contaminant;
Minimum distance to a populated area.

3.1.8 Temporal criteria

Can express the *persistence* of the effect;
Also the *length of time* considered for the effect.

3.1.9 Diverse areas covered by criteria

Since criteria are used to evaluate alternatives it is important to consider all areas of impacts related with these undertakings. A project such as the construction of an aluminum plant, for instance, will have of course criteria related with the economic aspect, but there are other areas equally as important as this one. Most probably the plant will be located on a seaside location for it will receive the raw material, alumina (aluminum oxide, Al_2O_3) from ships. This will probably involve construction of a harbor, with industrial piers, rail tracks and bulky equipment needed for offloading of the mineral. This will disturb the natural habitat of fish species, producing perhaps, because of the piers, some sand shifts.

Besides, the vegetation around the plant will probably be affected because of gas emissions, therefore the environmental aspect has to be addressed. Furthermore, and most probably, the plant will hire personnel from the area or lure them from other locations. This involves the construction of houses, the provision of services, amenities, recreation centers, etc., and certainly it will cause an impact in the fabric of the society. Needless to say, the damaging effects on the population by the chemicals released and the gases spewed have to be considered even after mitigation measures have taken place.

A very important component frequently overlooked is the production of odors. For instance a large meatpacking plant, or a sewage treatment plant can create pervasive odors which will be the source of future problems for the population nearby the plants.

From these few examples it is clear that a very important aspect is the *identification of criteria* to be used, which is done in a process (as noted above) called *scoping*.

3.1.10 Criteria weight

Usually a weight is assigned to each criterion in a percentage scale. There are different ways to obtain criteria weights. Most of them utilize *expert opinion* that gauges the comparative importance of one criterion against the others. This is the method used by the Analytical Hierarchy Process (AHP) (see section 5.8). All these methods involve *subjectivity* in the determination of criteria weights, and, which is probably more important, they *cannot be replicated* by another set of experts. For this reason Professor Milan Zeleny, an American mathematician, (see Bibliography - Zeleny M.), devised a very elegant method to determine criteria weights in which there is *no subjectivity*, since the model extracts the weights from the coefficients of the performance matrix (see sections 5.8.1 and A.3.1), which admittedly are most subjective.

Considering several alternatives and a set of say five criteria to appraise them, alternatives can be arranged in columns and criteria in rows. There is one criterion for each row, so one criterion is used to assess several alternatives. Each alternative has been assigned a coefficient which considers how well each option meets the objective regarding this particular criteria, and generally these values are normalized, being less than "1" and greater than "0".

There could be a large discrepancy in these coefficients for each criterion. The greater the discrepancy the better because it means that the corresponding criterion can *discriminate* or differentiate between alternatives. To measure these degrees of discrimination Zeleny employs a concept developed by the

American mathematician Claude E. Shannon, in his "Mathematical Theory of Communication", in 1948 (see Bibliography - Shannon C.). In this very famous paper which is considered the origin of Information Theory, Shannon established a measure of the *information content* in a message, that he called *entropy,* which is a very well-known function in thermodynamics, where it measures the level of disorder.

Shannon's formula is:

$$\text{H(entropy)} = -K \sum_{1}^{n} pi \log pi$$

K = Constant, which depends on the choice of the unit of measure
p_i = Probability of an event occurring.

Zeleny applied this concept but utilizing the coefficient values instead of the probabilities, and so, using this formula for each criterion, he determined which of them provides the *maximum quantity of information*, which is a measure of its importance of weight. The author of this book has also applied the same entropy concept but with the purpose of *selecting the alternatives that offer the most information,* when the system is applied to the determination of indicators (see section 6.8).

There is however another method which can determine criteria weights without subjectivity. This is called Mathematical Programming (see section 5.9), and this information comes automatically when solving the problem. In this case, the values assigned to each criterion represent the imputed values to them, and also correspond to the *shadow prices* (see Appendix section A.7).

3.1.11 Criteria thresholds

A threshold is a number which is used *to limit or set bounds* for a criterion. In this way thresholds are used as *yardsticks* to indicate if a project or an alternative has a significant effect. Many criteria use thresholds corresponding to indicators. For example if a criterion is related with the water consumption per capita in a city, it can be established at a value of about 255 liters/person which is the international standard. When this criterion is then used to evaluate a project contribution to reach a certain objective, this value is taken into account.

By the same token a threshold can also represent some measure of *sustainability* such as the rate of the recharge for an aquifer. This value can then indicate the *carrying capacity of the environment*, that is they inform on the capacity of the environment to sustain life.

Examples of threshold units are detailed in Table 3.1.

Table 3.1 **Examples of threshold units**

Area	Units
Environment	
Urban solid waste	kg / person-day
Maximum content of CO in streets in 8 hour period	9 µg / m^3
Paper recycling	%
Particles in suspension	mg / m^3
Infrastructure	
Street flooded with heavy rain	%
Dwellings connected to drinking water	%
Traffic flow	vehicles / hour
Transportation	
Expenditure in road infrastructure	$ / capita
Social	
Households below the poverty line	%
Total number of housing units	houses / 1000 persons
Median usable living space per person	m^2
Government	
Wages of local government staff to local expenditures	%

3.1.12 Threshold standards

Thresholds are assigned to criteria in most of the applications, and can be applied to any type of criterion. Examples are:

- Values for maximum capacity in utility plants such as water treatment and waste water treatment plants;
- Landfill capacities;
- Water, air and soil contamination indicators;
- Risk values;
- Urban indicators;
- Municipal service capacities, such as hospitals, police;
- Etc.

These threshold standards are then used to compare with values of impacts produced by projects and alternatives, and are also very useful in the monitoring stage. There is a good example of the use of these indicators in: *Monitoring Report Technological and Environmental Management Network Ltd January 24th 2002. Dredging and reclamation programme in Kingston Harbour*
Address:
http://www.nrca.org/publications/coastal/monitoring_rep/reports/MONITORINGREPORT4.htm

In this publication it is shown how the *actual values* emerging from the dredging operations *compare* with the *thresholds standards* established for each criterion.

As an example Table 3.2 (see Bibliography---Seinfeld J.H and Pandis, 1998), indicates the concentration of some gases in clean air, which can be used as standards or benchmarks.

Table 3. 2 **Concentration in clean air**

Gases	Concentration in ppb
Sulphur Dioxide	1-10
Carbon Monoxide	120
Nitrogen Monoxide	0.01 - 0.05
Nitrogen Dioxide	0.1 - 0.5
Photochemical Oxidants (Ozone)	20 - 80
Nitric Acid	0.02 - 0.3
Ammonia	1
Formaldehyde	0.4

Source: Seinfeld, J.H. Atmospheric Chemistry and Physics of Air Pollution, J. Wiley & Sons, 1986.
Reprinted with kind permission of John Seinfeld.

Table 3.3 points to the main origin of some contaminants.

Table 3.3 **Main origin of some pollutants**

Pollutants	Chemical symbol	Mainly originated from
Nitrogen gases	NOx	- Cars, thermal power plants.
Sulphur gases	SOx	- Coal fired power plants using coal with a high content of sulfur. - Oil refineries.

Carbon Monoxide	CO	- Produced mainly by cars especially when idling.
	Hydrocarbons	- Vehicles operated with gas and diesel oil. - Animal farms, producing methane.
Several gases	Smog	- Mainly from vehicles combined with solar action.
	Particulate	- Coal fired power plants. - Industries.
	Dust	- Construction works. - Atmospheric action.
Lead	Pb	- Vehicles using leaded gasoline.

3.1.13 Criteria relationships

Sometimes criteria are *related with each other*. In this circumstance it could very well be that some of them are redundant. Factor Analysis (FA) can be used to eliminate this redundancy. See section 3.2.

3.1.14 Assessment of significance

This very important and little- known concept tests for a certain criterion, which is the *part of the damage left* (residual effects) after mitigation measures take effect.

3.1.15 Criteria selection

How are criteria selected?

This is a question with no simple answer. Criteria selection depends on many features, such as type of alternatives, areas that probably will be affected for each alternative, importance of the project, data availability, etc. So, it will probably be safe to consider as many criteria as possible to be sure that everything is covered. Well, not quite, first because of the computational work involved, to say nothing about the monumental data gathering needed. In the second place because it is supposed that one criterion is not related with another because that could lead to a double counting of effects. How do we then proceed? As mentioned before, it is possible to use a technique known as Factor Analysis.

3.2 Factor Analysis (FA)

This is a statistical technique employed to *reduce data* and for discovering the *underlying structure* for a set of data. Let's explain first data reduction. Assume that for an EIA a set of criteria have been selected regarding road alternatives. Suppose that one criterion refers to "number of vehicles/hr on the road", a second criterion is "vehicle speed", a third criterion is "road width", a fourth criterion is "cold weather", and a fifth criterion is one such as "road safety" . So the five criteria are considered as well as many others related with the gauging of a set of alternatives or projects.

When the criteria were chosen probably they were selected considering their importance related with the alternatives, and this is fine. However, are all of these criteria independent one of another? At first glance it appears that they are not, but we don't know that for sure. Why is there interest in this question?

Well, firstly because the larger the number of criteria the more complicated the problem, and secondly since if they are somehow connected, then possibly *one effect will be counted twice.* From here it follows that it would be convenient to be certain that the different criteria are not linked between themselves. This can be done using Factor Analysis. Suppose that we analyzed the correlation between "vehicle speed" and "road safety", and we get a cluster or cloud of points, (or scatter plot), corresponding to these two variables, that is to say we can have several values for different speeds, and the corresponding values for accidents. See Figure A.3 in Appendix, section A.1.2.

Most possible this cloud of points will have a high correlation, say about 89 per cent, which means that the cloud will be narrow and that even it can be visually seen that it is possible to draw by hand a regression line (see Appendix, section A.1.1), with most of the points lying very close to this regression line and at both sides of it. If this is so, and having the equation of the regression line, then we don't need the two variables, since knowing one of them, say "vehicle speed", we can determine very accurately the other, and because of this the two variables have been reduced to only one.

This new variable is called a *factor;* it is a linear combination of the two intervening variables, can explain how they are correlated, and with it *redundancy* of information is avoided.

Let's now talk about the second characteristic, the discovering of an underlying structure. In the example proposed, why are accidents produced? Because the road is narrow? Well, obviously it appears sensible to assume that in a two lane road accidents occur, most probably, because a driver accelerates, tries to overpass another car and hits head on with a vehicle coming from the opposite direction. This accident can result from poor judgment, nervousness, anxiety to reach the destination, tiredness, etc.

Now, why do accidents also occur in a six lane highway with a separation rail from the other six lanes, on a beautiful summer day and with unlimited visibility? Certainly not because the highway is narrow. In other words there is a variable called a *latent variable* that cannot be seen, but its existence is inferred because it is possible to observe its consequences. It was discovered some time ago that this hidden variable exists, even if it is not possible to see it, and it is called *monotony, boredom, or tedium.* Truck drivers know it well since on a long and straight highway they get sleepy and accidents occurs.

This is the same underlying variable that originated the installation of a safety device in diesel electric locomotives; it is a pedal that has to be continuously depressed, and if the engineer falls asleep, as he relaxes, the pressure on the pedal diminishes and then an automatic mechanism stops the train.

It is also the same problem encountered in long flights when both the pilot and the copilot, because of boredom, and also because the plane is under automatic pilot, can involuntarily close their eyes and sleep. So, these two characteristics, that is, data reduction and detecting an underlying structure characterize Factor Analysis. Where can this new concept be used in EIA?

As it was said it will reduce the number of criteria to be considered and will determine the existence of the underlying structure. As a consequence, the result will show criteria as factors which are definitely not related one to another, so *no double counting can occur.*

There is another valued advantage. In analyzing alternatives there can be many different criteria regarding several kind of impacts in the economic, environmental and social field, forming groups or clusters. FA can also be used to detect these *clusters* of information, that is, data regarding economic conditions, another for social aspects, other for environment, and so on; in other words FA can tell us how many different fields or clusters we are contemplating in our selection of criteria. And why is this important?

Because then it is possible to know how balanced is the criteria selection, i.e., it allows us to know if the criteria are favoring some field, such as for instance giving more importance to environmental impacts than to social impacts.

3.3 Steps for Environmental Impact Assessment

The widely accepted steps for EIA implementation are:

3.3.1 Screening

This is the process that *identifies the source or origins* of impacts, in other words which are the *stressors* of the project. It also determines if a project, because of its nature and/or size, requires an EIA.

3.3.2 Scoping

Scoping *identifies the most important impacts* caused by the project. This stage is perhaps one of the most difficult since it has to make an inventory of issues that can be affected by the project. Scoping looks in the past, the present and the future, in relation to elements that are connected with the project to be developed. It also determines the alternatives, criteria to be considered, mitigation measures and residual impacts. Usually *indicators* are incorporated to analyze and measure the impacts.

The studies or analysis that have been performed prior to the present moment are not included here. It rather applies to the determination of the cumulative effects created by the elements intervening in the project. For instance if a project consists in the construction of a very large bridge over a river, it will probably be supported on huge concrete pillars, and then an anlaysis has to consider the effects produced by their construction in what is called a backward and present analysis.

Backward analysis

Tries to determine the contribution to pollution generated by the production of portland cement, reinforced steel, and other elements that will be direct inputs to the construction of the pillars.

Present analysis

Materializes during the construction stage. In this step one considers environmental perturbation because the construction of cofferdams to excavate the river bed to lay the pillar foundations, or perhaps to divert the river from its original course. There is a lot of earth moving in this operation, a large quantity of dust generated, and a high volume of noise, to say nothing of the fumes produced by the construction equipment. There is no doubt that these actions and respective consequences will perturb the wild life in the area and most probably the annual spawning of certain fishes that go upstream in the river for this function.

Forward analysis

Refers to the pollution produced as a consequence of the utilization of the bridge. Accordingly, it includes the emission of pollutants by the vehicles using it. There probably also will be a modification of the riverbed regarding the movement and/or erosion of the soil around the base of each pillar. The future analysis is also related with the dismantling of the bridge at the end of its useful life.

3.3.3 Remediation

This is a word normally used in EIA and indicates works that have to be done in order to eliminate or at least ameliorate the damage produced to the environment. Take as an example the construction of roads connecting a bridge construction site with a rock quarry, and consider that these roads will be useless when the bridge structure is finished. In this case *remediation work* includes the *restoration* of the disturbed soil to its primitive condition as close as possible or still better, improving it. For instance, something has to be done with the quarry. There is a very interesting example of remediation in the City of Vancouver, Canada, where an old quarry was converted into a beautiful garden.

Another example was the demolition of an old oil refinery in the heart of Mexico City. In this case for instance, once the old equipment was dismantled, the soil had to be restored because it was highly contaminated with metals, chemicals and oil. The resulting restoration produced a park in an area of the city which was lacking this amenity. Sometimes mitigation is not related with restoration. For instance the sound barrier built for a highway project is not to restore the environment to its pristine condition, but to *diminish* the level of noise produced by the traffic on the highway.

3.3.4 Impact assessment

Once it is known which areas will be affected by the development, it is necessary to have an estimate of the benefits and damages. Another very difficult task indeed!

Normally in an EIA one works with a panel of experts, who determine not only the criteria to be employed but also estimate how much each alternative contributes to each criterion. This is usually done using the Delphi method (see section 4.3.7), and a *pair-wise comparison* between different alternatives and different criteria (see section 5.8). Expert opinion also *establishes the weight* to be assigned to each criterion relative to the others (see section 5.8).

In this way, a matrix or table can be constructed having alternatives or projects in columns and criteria in rows. Each cell in the intersection of a column with a row is assigned a value as per indicated above. It is also necessary to establish thresholds for each criterion, as commented in section 3.1.11.

3.3.5 Impact mitigation

As explained above it is necessary to have data about the *environmental damage* created by the project and *evaluate the necessary mitigation measures* as well as *restoration actions.* Impact mitigation depends definitely on the type of impact and related project. It is possible for instance to mitigate the air contamination produced in a city for vehicles just by limiting the volume of cars entering a zone of the city, typically the CBD (City Business District).

Another mitigation measure is the already mentioned construction of sound barriers. Another could be the planting of trees to replace those logged by the project. In the construction of an airport some sound mitigation could involve enforcing new regulations pertaining to take-offs.

3.3.6 Public participation

There has been a trend recently to let people affected by a project express their opinions, and to suggest solutions. This is very important and possibly does not have the attention it deserves. People who live in the area are the best suited to point out problems that decision-makers and staff do not see. For example, in a European city an urban highway was designed according to economic and engineering principles, albeit the social part had not been taken into account. Fortunately, before the project was approved, a survey showed that many people were angry with the plan because it would, for several kilometers, prohibit access to a park along a river that attracted thousands of people from around the area. If the highway went ahead as designed, the people living on the other side of the highway would see their access to the park cut.

The answer, although not cheap, was simple, and implied the construction of several underpasses below the highway. This solution also diminished the cost of a scheme to drain rain water from one side of the highway to the river, by constructing --- together with the underpasses ---storm water drains. There are many ways for people to be included in a participatory process. For instance, it is possible to conduct surveys and polls, to organize meetings with representatives of the local population, to have these representatives serve as the link between citizens and the decision-makers, etc.

Usually surveys are conducted using a questionnaire containing a number of queries and requesting citizens to answer. This questionnaire can be mailed or filled out in personal direct interviews. Of course, it is physically impossible, let alone expensive, to interview hundreds or thousands of people, unless a full, formal census is taken. In most cases this is not economically feasible.

As a consequence, it is necessary to operate with *samples,* which allow one to work with a small, representative, number of people, and from them infer results about the whole population. Needless to say, the information thus gathered is subject to many constraints and assumptions, and is valid only within certain limits, but nevertheless the Theory of Sampling, which lies in the field of Statistics, is a proven tool and used very frequently. Remember that the pre-election polls usually forecast results within perhaps a 5 % margin of error. The same tool can be used here.

The questionnaire itself has to be prepared very carefully. It cannot be too short because in that case the answers will be very constrained. It cannot be very long because people get tired or frustrated with long questionnaires and then give any answer just to get rid of the interviewer. The questionnaire within its limitations has to have a broad scope to cover the different aspects which are of interest. So, a questionnaire about the construction of an urban highway for instance has to cover queries including social, economic and environmental issues.

Also important is to have an equilibrated sample, that is including a cross-section of population with diverse educational and social levels, as well as with different economic means.

Two ways to facilitate people participation are now discussed, and they apply to both, proposals from the citizens and proposal from City Hall. These two methods have been proved effective in their urban jurisdictions. The first one we called "People Sharing" and the second one "People Participating".

"People Sharing" is more in line with continuous projects in a city and works in this manner:

1. City Hall establishes a program under which each area of the city selects a representative from its own people.

2. These representatives are trained at City Hall expense in several issues such as: preparing proposals, screening demands from people, gathering information, etc. The idea is to have a team *well aware of the needs, responses and wishes of the community they represent.* The team is then able to prepare different plans, that respond to actual needs from people, and to submit them to the respective management level in City Hall.

   For the same token, if a project originated in City Hall could have negative impacts, citizens are given through their representatives, an opportunity to express their opinions regarding different aspects of the proposal.

   City Hall receives these proposals, comments, and suggestions which have been screened by each representative, and then sends technical people to *discuss with the neighbors the technical aspects involved in each one.* In these meetings these City Hall officers receive feedback from the citizens and modify the proposals according to this input.

3. City Hall possesses then a listing of projects, which are *feasible,* and that *represent the wishes and needs of the people,* or it can have *a listing of the main subjects or objections posed by the people on a certain proposal.*

4. The final stage involves the *selection of projects to be executed* considering monetary restrictions as well as the use of other scarce resources, or the modification of certain aspects of the City Hall proposal and in accordance with input received from the citizens.

The second method "People Participating" works in a different manner and is more appropriate to individual projects and asking for people's opinions. In this method, City Hall organizes *medium size meetings where citizens as well as stakeholders can express their opinions*. In this setting each meeting is chaired by a moderator.

For a specific project people are requested to make their feelings known, as well as to express doubts, concerns, and ideas about anything that is related with the project. This is done through a procedure where participants can express anonymously --- via computers installed in the meeting area and a dedicated software --- what they think about the project. Their thoughts are reflected on a large screen for everybody to see. The anonymously expressed concepts are discussed publicly with all the attendees, with the moderator consolidating those issues that are similar but that have been presented in different manners.

After the discussions are over the attendees are requested to vote, again, anonymously, about each subject or proposed option. The number of votes that each proposal or subject receive is automatically weighted and ranked by the software. The system can be complemented by opinions from people not present in the meeting, through their personal computers. These ideas are then incorporated into the others discussed in the meeting.

Once the meeting is over, the conclusions are printed and mailed to the attendees and whoever wishes to receive them. In this way there will be a larger probability that comments from a representative sector of the population affected by the project will be considered in the project appraisal.

Internet references for Chapter 3

CONCEPT: Significance
Title: *Specific methodologies and criteria to determine adverse transboundary impact*
Develops a Table with the identification of adverse transboundary impacts for:

- Air;
- Water;
- Climate;
- Soil;
- Landscape/Historic monuments or other physical structures;
- Human health and safety;
- Flora and fauna.

Table provides elements for the determination of the significance of impacts.

There is also a questionnaire on past experience with transboundary impacts http://www.unece.org/env/eia/cepwg3r6.htm

CONCEPT : Cumulative environmental effects - Hydroelectric
Title: *Methodology for assessing the cumulative environmental effects of hydroelectric development of fish and wildlife in the Columbia River Basin*
Extensive document (150 pages), considering different kinds of impacts.
http://www.efw.bpa.gov/Environment/EW/EWP/DOCS/REPORTS/GENERAL/I19461-3.pdf

CONCEPT: Cumulative effects
Title: *Considering Cumulative Effects Under the National Environmental Policy Act*
Paper produced by the National Environmental Policy Act (NEPA), USA.
Includes:

1. Introduction to cumulative effects analysis;
2. Scoping for cumulative effects;
3. Describing the affected environment;
4. Determining the environmental consequences of cumulative effects;
5. Methods, techniques and tools for analyzing cumulative effects;
6. Examples of each methodology.

http://ceq.eh.doe.gov/nepa/ccenepa/ccenepa.htm

CONCEPT: Thresholds
Title: *Thresholds of significance*
State of California - Governor's Office of Planning and Research.

According to the paper, *"Thresholds of Significance" discusses how public agencies, including cities, counties, and special districts, may adopt quantitative or qualitative thresholds which represent the point at which a given environmental effect will be considered significant. Enacting thresholds helps ensure that during the initial study phase of environmental review, significance determinations will be made on a consistent and objective basis.*

The paper analyses the following important concepts:

1. Advantages of adopting thresholds;
2. Establishing thresholds;
3. Drafting thresholds;
4. Tips for thresholds;
5. Limitations.

http://ceres.ca.gov/topic/env_law/ceqa/more/tas/Threshold.html

CONCEPT: Scoping
Title: *Guidance on EIA - Scoping*
Environmental Resource Management
Very complete and comprehensive information on scoping in this 35 pages report.
http://www.dhi.dk/Courses/AlumniCafe/LectureNotes/EU%20guide%20on%20scoping.pdf

CONCEPT: Scoping
Title: *Scoping Guidelines for the Environmental Impact Assessment of Projects*
Environmental Agency, Government of U.K.
Very useful document. It can be requested at:
commercial.policy.unit@environment-agency.gov.uk
Address of this reference:
http://www.environment-agency.gov.uk/business/444304/502508/311024/?version=1&lang=_e

PART II: ELEMENTS OF ENVIRONMENTAL APPRAISAL

CHAPTER 4 - INFORMATION FOR EIA

Part II of this book is related with two main aspects of this process: a) the collection of information and b) the processing of this data. For the first part of Chapter 4, different models for gathering and analyzing information are described and illustrated with case studies. The second part is discussed in Chapter 5, also with case studies demonstrating each model.

However, there is no clear distinction between the two sets of techniques, therefore in some cases gathering information can give enough data to make a decision, especially in simple scenarios.

4.1 Defining the problem

A procedure as complex as an EIA involves the participation of many people at different levels. This is so due to the different disciplines involved and the various degrees of expertise required. Because of the existence of diverse circumstances there are also varied methods for gathering data and its analysis. The next section explores the tools available to the analysts.

4.2 Available tools for project appraisal

An EIA tries to identify *trade-offs*, from alternatives or options conforming to an established objective. These different alternatives are gauged against a set of criteria pertaining to different fields.

If the whole system is considered together, that is projects, objectives and criteria, there are *gains and losses* materialized by trade-offs between these components of the system. As a consequence it is desirable to have a tool with the analytical power to *illustrate these trade-offs to allow people to make a decision.* This is not an easy task since it is almost impossible to keep track of all the gains and loses and to select the project that produces the best result, and for this reason it is necessary to develop a method that can help in this clarification, albeit, of course, the *final decision always is made by the decision- makers.*

An EIA attempts to identify alternatives, hopefully relying on a complete knowledge of their trade-offs. The question arises as to *how to inter-relate various components and how integration* can be achieved. Integration calls for information on economic, environmental, social, biophysical issues, citizens involvement, and sustainability when selecting an alternative; a crucial activity in that sequence is the *identification of alternative solutions* that take into account cross-sectional issues.

4.3 Tools for impact identification

There are at the present time a set of tools that are used to make this identification, as follows:

4.3.1 Checklists

Description

A checklist, as its name implies, is a listing of potential impacts that a project may create. As a consequence, a project has to be analyzed considering not only the environmental factor --- and not t related only to the environment --- but other fields as well, such as social and ecosystems, considering its beneficial or adverse impacts on them. For instance:

- Population to be affected by the project;
- Soil, air and water;
- Flora and fauna;
- Land use;
- Wilderness;
- Potential risks;
- Etc.

The checklist needs to be organized for the type of project under consideration, and usually is prepared in a form involving a series of questions. A case study will exemplify the preparation of such a list.

4.3.1.1 Case study - Complex for energy generation and transmission

Background information

To illustrate a checklist preparation, this case study involves the construction of an underground facility designed to generate hydro-electricity using the level difference between a lake (Upper Lake), high in the mountains and the Bear River, in a valley, with a total head (level difference between the water source and the river), of 241 meters.

The powerhouse was to be built underground, the same as the penstock (the pipeline conducting water from the lake to the turbines). The electricity generated was to be transported using a high voltage power-line, through the wilderness to a city located 76 km away. This wilderness is a very beautiful area, with mighty mountains, rich forests, plenty of wildlife, abundant fishing, and with very cold weather.

So, the project will be in the middle of a very sensitive area; however, because it is designed as an underground power station, there will be no visual components, other than the power-line. So, it is believed --- and computer simulation has proved it --- that the wilderness beauty will not be aesthetically affected. However, the area is populated by black and grizzly bears, which would pose some danger for the 750 people lodged in a construction camp. Also, because of this human population, waste and waste water will be generated, in the amount of about 200,000 liters a day.

There is no doubt that this project would generate impacts of every kind in the environment, so an environmental protection plan was designed and aimed to be strictly enforced. This project will be utilized to prepare an example of a checklist for environmental impacts. Several sources of impacts are considered here because the project involved the following *areas which are spatially distant:*

- Upper Lake;
- Main Camp;
- Penstock;
- Caverns for the turbines and transformers;
- Transmission lines.

Table 4.1 shows the checklist for this project.

Table 4.1 **Checklist for the hydro-power project considering only the environmental point of view**

Areas	**IMPACTS**	**YES**	**NO**	**Not enough data**
Upper Lake	**Requirements and issues at this location**			
	Installation of a construction camp for more than 50 people?	x		

	Installation of an incinerator?	x		
	Installation of a water treatment plant for sewage?	x		
	Earth moving?		x	
	Logging to make room for the camp?	x		
	Will it affect wildlife?			x
	Will there be a pollution risk because of treated water discharged into the lake?	x		
	Will there be a considerable volume of fuel in the area?	x		
	Will there be a considerable truck traffic?		x	
	Will explosions produce avalanches?		x	
	Will there be difference in the lake water level because the water piped out of it?	x		
	Will the ecosystem be affected?			x
	Will be the fish population be affected by the water intake in the lake?	x		
Main Camp	**Requirements and issues at this location**			
	Installation of a construction camp for more than 50 people?	x		
	Installation of an incinerator?	x		
	Earth moving?	x		
	Logging to make room for the camp?	x		
	Will it affect wildlife?	x		
	Will there be a pollution risk because of treated water discharged into the river?		x	
	Will there be a considerable volume of fuel in the area?	x		
	Will there be a heavy truck traffic?	x		
	Need for paving?	x		
	Will explosions produce avalanches?	x		
Excavation for the penstock	**At this location**			
	Will explosions for tunneling affect wildlife because of noise or danger to species?		x	
	Will removal of blasted rock require heavy truck traffic in the area?	x		
	Will the dumping of blasted rock affect the ecology of the area?	x		
	Will there be chemicals products impacting the environment from this blasting?	x		
	Will there be exhaust gases from explosions?	x		

Caverns	**At this location**			
	Will explosions affect wildlife because of noise or danger to species?		x	
	Will removal of blasted rock require heavy truck traffic in the area?	x		
	Will the dumping of blasted rock affect the ecology of the area?	x		
	Will there be chemical products impacting the environment from this blasting?	x		
	Will there be concrete poured in large quantities?	x		
	Will there be exhaust gases from explosions?	x		
Transmission line	**At this location**			
	Will explosions affect wildlife because of noise or danger to species?		x	
	Will the dumping of blasted rock affect the ecology of the area?	x		
	Will there be chemicalsproducts impacting the environment from this blasting?		x	
	Will there be concrete poured in large quantities?	x		
	Will there be exhaust gases from explosions?		x	

A weight could have been added to each one of the impact components and even another weight assigned to each component, using for instance a 1-10 scale.

Mitigation measures

Several mitigation measures have been designed to address some specific impacts, as shown in Table 4.2. This is only an example and does not cover all the impacts that a project of this magnitude can produce. Besides, notice that we consider here only the *environmental aspect of the project* with a very little incursion into the *social component* (since the installation of the construction camp includes board and lodging, recreation and sport facilities for workers, etc.). Naturally, all the remediation measures than could have been taken are not mentioned here, but of course, in a real life case they should be considered.

Table 4.2 **Mitigation measures taken in the hydro-power project**
(Only Upper Lake and Main Camp considered)

Areas		
	Impact	**Remediation and restoring**
Upper Lake		
	Installation of a construction camp for more than 50 people.	At the conclusion of the work all constructions will be removed, and the soil restored to its original conditions.
	Installation of an incinerator.	The incinerator will be discontinued, and removed from the site.
	Installation of a water treatment plant for sewage.	The water treatment plant for sewage is buried and will be closed.
	Earth moving.	Soil will be restored to its original conditions.
	Will the dumping of blasted rock affect the ecology of the area.	The effect of blasted rock dumped in the area will be greatly reduced because that rock will be crushed and used for concrete production.
	Logging to make room for the Camp.	Trees will be planted in the area used for the Camp.
	There will be a considerable volume of fuel in the area.	It is intended to keep a close track of all damaging spills. In cases where action was not taken, the soil will be treated to get rid of hydrocarbons.
	There will be a difference in the lake water level because the water piped out of it.	There is no solution for this problem. There will be a residual impact which can translate in less water going to the river that discharges the lake.
	Will the fish population be affected by the water intake in the lake.	If nothing is done many fishes will be sucked in by the water intake and destroyed . This is not only an environmental problem but also a technical one, since the accumulation of fishes can seal off the entrance of water to the water nozzles. The solution is to build the intake with a metallic grate as well as a mechanical cleaning mechanism.
Main Camp		
	Installation of a construction camp for more than 50 people.	At the conclusion of the work all constructions will be removed, and the soil restored at its original conditions.
	Installation of an incinerator.	The incinerator will be kept since it will continue to serve the population in the village serving the plant, but of course, the emissions will be only a fraction of what it was during the construction period. It is assumed that the surrounding forest can absorb most of the emissions.
	Disturbed soil.	Soil disturbed will be restored to original condition, included trees.

	Logging to make room for the camp.	Trees will be planted in the area used for the camp.
	Will it affect wildlife.	Bears in the Main Camp area will be transported to other sites to avoid incidents with people. At the conclusion of the project these animals will be transported again to their original habitat.
	There will be a considerable volume of fuel in the area.	It is intended to keep a close track of all damaging spills. In cases where action was not taken, the soil will be treated to get rid of hydrocarbons.
	Need for paving.	Paving in non-essential areas will be removed, to restore soil permeability.

What is missing in this check list for environmental effects?
It does not mention secondary effects that can take place as a consequence of a direct effect, and *this is a drawback of checklists*. What secondary effects could they be?

For starters, extracting water from the Upper Lake through the penstock will produce a decrease in the water level, which is a *direct effect*. The level reduction will cause the exposure of the sandy beach in the lake shoreline, and because they will be without vegetal protection, there is a probability of erosion as a *secondary effect*. This in turn can cause the water to erode the banks of the lake until it reaches the tree level, producing the falling of existing trees into the lake, which can be considered as a *tertiary effect*.

On the other hand, there are two wetland areas nearby. These areas will drain into the lake when its level drops, causing their shrinking or perhaps drying up, and posing danger for wildlife (*another secondary effect*). It is realized that now there will be *cumulative effects* in the whole area, considering erosion and draining of the wetland, which can be significant when considered jointly.

Besides, kilometers away and 241 meters below the lake level runs the Bear River, where the water from the Upper Lake will be discharged after producing work in the hydro-turbines. This is a fast mountain river whose flow greatly increases during Spring with the melting of the snow in the mountains. Once the project is in operation, it will discharge an amount of water which is roughly equivalent to the actual flow of the river. As a consequence, especially in Spring, the river will possibly overflow and low areas along its course will be flooded.

It can then be seen how an impact in the source area (the lowering of the water level), produces some secondary and tertiary effects not only in the source area --- which also becomes the receptor area --- but also in the valley of the Bear River, *spatially located* many kilometers away. A *temporal effect* can also be envisaged due to the above mentioned erosion of the lake's shoreline, which can take place *sometime in the future* once the project has been completed.

A project very similar to this example was actually under construction in the 1990's in Western Canada. After years of work and hundreds of million of dollars invested the project was definitely halted. Why?

Because native people, living on the shores of a river fed by a lake similar to the Upper Lake of the example, argued that the project was diminishing the stock of fish that were their major source of food. They sued the company executing the project, brought it to the Supreme Court of Canada and won. Without pretending to analyze who was right and who was wrong, this actual case shows very clearly that in *projects spatially dispersed* such as in this Canadian project, where the source of the impact (the lack of enough water in the river discharging the lake) was hundreds of kilometers away from the receptor of the benefits, (the city where the electricity was to be sent), *the spatial component of the impact has a very important weight.*

That also shows how the benefits that the generated energy could have brought to the city in better living, creation of industries, etc., also caused harm to people at the source because their lifestyle was to be degraded.

Conclusion

In this actual case, a secondary effect --- the lack of water in the river draining the lake --- was argued as cause for a reduction in the fish population, which was strong enough to cancel this more than a billion dollar project.

4.3.2 Network analysis

Description

In this system a *cause and effect relationship* is detected analyzing different areas likely to be affected by the project. Usually it is performed utilizing a block diagram, that is boxes linked with arrows, indicating which are the *stressors* and which the *receptors.* A block diagram or network can be used and the arrows indicating the relationships of the different actions of a project with their corresponding impacts. Because this block diagram shows *a sequence of results it can depict for an action the primary, secondary, etc.*

impacts it produces on a receptor, and it can also show impacts from other actions outside this sequence on the same receptor. See Figure 4.1.

The solid arrows show first impact and broken line arrows indicate secondary effects. For instance the construction of a dam can hinder the yearly swimming upstream of fish for spawning (consequence 1), *direct effect.* Also, and because the river is no longer depositing rich sediment, the fish population in the mouth of the river is decreasing (consequence 2). Further, and because of both impacts, the canning industry, located at the mouth of the river is not getting enough catches, *secondary effect.* Of course, we can follow the chain of events and see that the decreasing of industrial activity in the canning plant produces layoffs, *third effect,* which in turn............

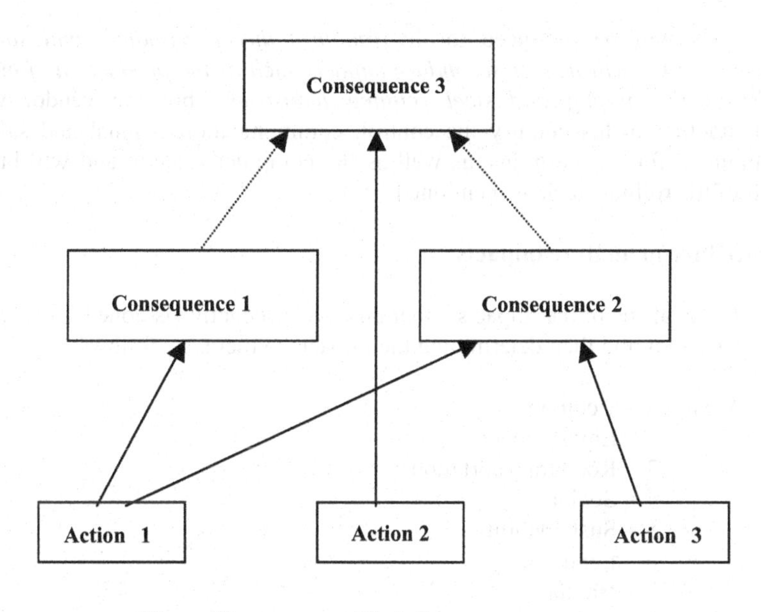

Figure 4.1. **Block diagram**

4.3.2.1 Case study - High speed train project

Background information

As an example of network analysis consider a case where a country plans to build a high speed passenger rail link between its capital city, with a population in excess of 5 million, and the second city in the country, with a population of about 2.3 million. The cities are separated by 482 km, and a train traveling at a cruise speed of 235 km/hr can cover the distance in 2.15 hr, including two stops in intermediate cities, and with a service of about 14 trains each way per day. This project represents a huge investment for the country, since it has to acquire the technology and import part of the equipment from a foreign origin.

The overseas vendor has agreed to establish a local plant for the manufacture of 157 cars and 18 locomotives -- out of a total of 314 cars and 37 locomotives -- that will be installed in a city roughly halfway between the two terminal cities *(there then will be direct economic and social impacts or consequences, which in turn create secondary and tertiary consequences).* The balance of equipment will be built in the vendor's country.

Rails will be produced locally *(another direct economic and social impact, which will trigger secondary impacts such as the production of more jobs in the mining and steel refining industries),* but the vendor will manufacture in his country the control, communications, signal and safety equipment for all the trains, as well as the computer system, and will build half of the rolling stock as mentioned.

Procedure to analyze impacts

The analysis of the impacts either positive or negative is done by *selecting several areas* and then determining the impacts to them, as follows:

Areas: 1 = Economy
2 = Environment
3 = Road transportation
4 = Social
5 = Sustainability
6 = Land use
7 = Habitat
8 = Safety

Economy area
Positive impact:
Communications:
The economy of the two cities will improve because the rail link will facilitate communications between both.

Negative impact:

Business:

The fast link will reduce the vehicular traffic which brought commercial benefits, derived from shopping, lodging and tourism in the area between the two terminal cities.

Positive impact:

Energy:

Because of the use of electric energy rather than fossil fuels, the country will benefit from reducing oil imports, since there will be a dedicated hydro-electric dam to be built for this project.

Positive impact

Productivity:

Given the reduction in travel time working hours for passengers will be more efficient.

Positive effect:

Air transportation:

Will practically come to an end because the air trip between downtown to downtown takes more or less the same time as the train and at a higher cost.

Negative effect:

Fare:

The cost of the ticket in the new system will be almost double that of the ticket in a bus, but as expected the travel time and the safety will be considerably improved.

Negative effect:

Balance of payment:

Because of the amount of equipment as well as the technology to be imported, the project will cause negative impact in the balance of payments.

Environment area

Positive impact:

Energy:

The new system will greatly reduce fossil fuels consumption since present diesel trains will be replaced by a high efficiency electrical system. On the other hand the large number of daily trips between the two cities (14 trains in each direction) will create a good ratio between passenger and energy consumption.

Positive impact:

Energy:

Because the number of tankers on the road to bring fossil fuel from the harbor will be strongly reduced, air pollution will greatly decrease.

Air pollution:

The new system will eliminate pollution caused by the exhausts of diesel trains.

Road transportation

Positive impact:

Energy:

Because of the use of electricity, truck traffic from the harbor to one of the cities will be reduced considerably, which means less fossil fuel consumption.

Social area

Negative impact:

Accessibility:

The link will decrease or maybe eliminate bus services between the two cities, so intermediate locations will be isolated.

Negative impact:

Relocation:

A large number of people will have to be relocated because their properties stand in the right of way for the tracks.

Positive impact:

Compensation:

People will be compensated not only by payments of the price of their land, but also by initiation of new commercial ventures.

Positive impact:

Employment:

By law some of the available job opportunities in the train operation will go to displaced people.

Positive impact:

Employment:

The project will create thousands of employment opportunities in the area for the manufacture of equipment, rail construction, and equipment maintenance.

Positive impact:

Travel time:

Will be drastically reduced from 6 hours to 2.15 hours.

Negative effect:

Air transportation:

Most probably the air carrier covering this route will be out of business, therefore there will be lay-offs of flying personnel.

Sustainability

Positive impact:

Renewable energy:

Its utilization will curtail the use of non-renewable resources.

Land use area

Negative impact:

Agricultural land :

The high speed demands that the tracks be as straight as possible, in order to avoid speed reduction in curves. For that reason the project will utilize agricultural land since it cannot make use of the existing right of way.

Habitat area
Negative impact:
Wildlife:
Will be disturbed by the noise produced by trains. On the other hand, habitat as it is now will be reduced because tracts of land will be used to build the elevated tracks.

Safety area
Positive impact:
Accidents:
Baseline conditions indicate that the highway linking the two cities is mostly a two lane road. At the present time and because of the intense traffic of trucks, buses and private cars, accidents are frequent, most of the time fatal. The new rail link will drastically curtail this rate of accidents, leaving the road mostly to trucks.

The above scenario can be condensed in a block diagram or in a matrix as shown in Table 4.3, although it is believed that the graphic representation gives a better idea of the interrelations. In Table 4.3 numbers in columns correspond to the areas detailed above. Different categories of impacts are depicted in rows. Only *direct effects* are considered but there is nothing to stop the introduction of *secondary, tertiary and more impacts*.

Letter "P" means a positive impact, while "N" is a negative one. The last column on the right indicates number of positive and negative or adverse impacts for the category. A number at the left of a letter indicates that for the same area and impact category there is more than one impact. In this example each area has been assigned an equal weight. The same applies for categories. However, it is believed that two things should be done:

1. Assign each area a relative weight, that can be derived perhaps using the AHP method. See section 5.8.

2. Compute for each impact category some degree of magnitude and importance.

With these two values, weight of areas and importance of impact categories, we will be in a much better position to assess the total impacts of this project.

Table 4.3 **Effects of impacts**

Areas	**1**	**2**	**3**	**4**	**5**	**6**	**7**	**8**	**P**	**N**
Impacts categories:										
Communications	**P**								**1**	
Business	**N**									**1**
Energy	**P**	**2P**	**P**						**4**	
Productivity	**P**								**1**	
Air transportation	**P**								**1**	
Fare	**N**									**1**
Balance of payments	**N**									**1**
Air pollution		**P**							**1**	
Accessibility				**N**						**1**
Relocation				**N**						**1**
Compensation				**P**					**1**	
Employment				**2P**					**2**	
Travel time				**P**					**1**	
Air transportation				**N**						**1**
Renewable energy					**P**				**1**	
Agricultural land						**N**				**1**
Wildlife							**N**			**1**
Accidents								**P**	**1**	

Conclusion

Network analysis provides an effective and visual way to illustrate cause-and-effect relationships. The block diagram is a very convenient method to show this relationship with the indication of positive or adverse impacts. It can also highlight when one action has more than one consequences, or when two or more consequences create a third one. The advantage of translating this information to a matrix is that then mathematical operations can be performed in a computer, such as the finding of the chain of events from actions to consequences. Also a severity indication of each action and subsequent consequence can be pointed out and a final figure for damage obtained.

4.3.3 Leopold's matrix.

Description

Developed by Dr. Luna Leopold of the United States Geological Survey, this is a table or matrix where in columns are listed the different actions or activities of a project (stressors), and in rows the environmental existent conditions that could be affected by that project (receptors). In the intersection of a column and a row there is a cell with two values representing the magnitude of the action and its importance. Both magnitude and importance can be expressed in a subjective way using a 1-10 scale. It provides a *convenient mean to show cause-and-effect relationships*, however it is not a mathematical matrix where true algebraic operations can be performed.

Since it uses subjective weights for relative importance of effects and impacts, *it is not objective*, and also *it does not establish interactions.* Besides, the decision-maker is faced with a table with perhaps hundreds of values which makes it difficult to select alternatives.

4.3.3.1 Case study - Construction of a water reservoir

Background information

As an example, assume that a project calls for the creation of a vast water reservoir to supply water to a large city, and through the construction of a dam. One of the actions or activities of this project will be the flooding of the area behind the proposed dam, which at the present time --- *baseline conditions* --- involves a small town, a forest and a road. If we say for instance that the flooding will have a negative magnitude impact of -8, it means that the rise of the waters will be of enough magnitude to submerge the town, the road, and cover part of the forest. A magnitude of -3 will perhaps produce the rising of the water level just a couple of meters without affecting the town. Its importance depends on how important are the town, the road and the forest.

From the economic --- and not from the humanitarian point of view --- perhaps the disappearance of the town does not have a great importance because people are *willing to be compensated* and relocated in other site. The road can be reconstructed without too much effort a couple of hundreds of meters from its actual location, and the forest, which is not a bird habitat, will suffer the diminishing of its forested area, but this can be remediated by replanting to compensate for the loss. So, as a bottom line perhaps the importance, always in the 1-10 scale, is only 5.

A similar actual scenario happened in the 1960's when the Aswan Dam was built. The magnitude of the flooding required to create the Nasser Lake was so large that it would submerge the Temples of Rameses II, dubbed the Eighth Wonder of the World. So, its magnitude was without a doubt equal to 10. Its importance was also probably equal to 10, because mankind would lose forever an invaluable archeological site. It was so significant, that a universal task force was assembled to relocate the Temples well above the rising waters of the Nasser Lake, where, fortunately, they stand today

In this context, the Leopold matrix is a very good pictorial device to show cause-and-effect relationship: *cause: **flooding,*** and *effect: **loss of artworks**.* Originally the Leopold matrix was prepared to include almost any action that a project can produce, and in so doing it involves 100 columns. Besides, it also covers a great deal of environmental conditions that can be affected by the project actions, and totaling 88 rows. Of course, it is unlikely that a project can have 100 actions with influence on 88 environmental conditions, so, in actual practice the only columns and rows used are those pertaining to a particular project.

Nevertheless, assuming that the project can have just 15 actions with impacts in 20 environmental conditions, there will be about 15 x 20 = 300 cells, each one with a set of two numbers, *magnitude* and *importance,* so the total number of values increases to 600. Everybody can realize that it is extremely difficult to handle this amount of information, and probably this is the main drawback of the model. However, it is believed that using multicriteria analysis, this difficulty can be solved.

With multicriteria analysis it is even possible to assign weights to the different criteria or environmental conditions, and still better, to assign thresholds to each one. In this case columns are *project actions* and rows can be considered as *criteria.*

Let's come back to the example of a water reservoir construction, where the following project actions are shown in Table 4.4.

Table 4.4 **Magnitudes and importance of impacts**
Project actions

		Excavation for the dam	**Flooding**	**Traffic**	**Con-struction**
Environmental impacts or conditions					
Human population in town	Magnitude Importance	+ 9 + 8	-9 5		- 5 4
Ecology	Magnitude Importance	- 8 6	-7 5	-7 3	
Fish spawning	Magnitude Importance				-9 5

In this way it is feasible to express the same concepts as in the Leopold matrix, but in a clearer manner, and most important, determine which is the most damaging action. Let us explain the values:

Human population in town

- For the town people the construction of the dam has a significant magnitude because it will provide jobs for two years, and this is very important due to the fact that there are few available jobs in this small town.

- Flooding will certainly alter their living for they will have to move to another place, but this importance is relative, since the new settlement is only 7 km away, and because the new houses will be better than their present day dwellings.

- Construction will have definitely a negative effect because of noise, dust generation and higher costs for food, however its importance is low because of the benefits provided by the undertaking.

Ecology

- The excavation of the dam will provoke negative effects to the environment due to dust production, and since it will be necessary to log a large part of the forest to make room for the excavation for the river diversion during construction. However, its importance is not very high since the forest does not have an important bird population, and also because measures will be taken to control dust, through daily wetting of the construction dirt roads.

- Of course, flooding has a negative effect for the reason that the creation of the lake will generate a micro climate in the area whose effects are unknown, however, its importance albeit high, is not severe considering the experience gathered from other sites.

- Traffic will be intensive. There will be a large production of noxious gases from the heavy construction trucks, but it is not important because of the direction of the prevailing winds in the area.

Fish spawning

This is a serious ecological problem since the dam will be a barrier for the fish to spawn upstream. It has a large order of magnitude for it will alter the ecological equilibrium and will endanger the fishing industry located downstream. However its importance is low because the project foresees the construction of a "fish ladder", for the fish to overcome the dam and this is in reality a *mitigation measure.*
It is possible now to apply some of the techniques described in Chapter 5 to determine the ranking of project actions when the impacts are considered. In this case, the number of values we have in the matrix really does not matter.

Conclusion

The Leopold matrix has a series of drawbacks, one of them for instance is the lack of consideration of *time-dependent effects.* However, it is an important tool because its preparation *involves a systematic review of all the aspects* of a project, and especially because it requires the analyst to think about *magnitude* and *importance* for each impact or condition.

4.3.4 Flow diagrams
Description

Are suitable to identify action-and-effect relationships, but are not advantageous when several alternatives are considered.

4.3.5 Battelle environmental evaluation system
Description

It employs relative weights for indicators, and because its use of value functions has a high objectivity. Interactions are not indicated.

4.3.6 McHarg system - Overlays
Description

It has its origin back in 1968 when Dr. Ian McHarg devised an ingenious system to spatially explore an area. *Thematic transparent maps* were developed for flora, fauna, geology, population, rivers, slopes, roads, agricultural land, etc. These maps were then placed on a glass table, one on top of another, forming layers of information about the zone. When an intense electric lamp was placed beneath the glass table, the light reaching the top layer indicated those areas free of impediments for the location of the project under study. As an example, assume that it is necessary to build a highway from east to west and across an area where there are population and other features that have to be considered. The problem is to find a route for the highway with the minimum disturbance of the existing characteristics. Figure 4.2 shows the use of *overlays* when it is necessary to know the spatial interaction of these different land uses. In this example the following characteristics are considered, each one in its corresponding map-layer.

- A map indicating the location of a natural forest;
- A map showing the area dedicated to agricultural land;
- A topography map indicating different elevations to the west of the study area;
- A road map linking a city with two nearby towns and from the city to the west.

These four maps or layers are then overlaid and a composite map is obtained, with Figure 4.2 showing the superimposing of the four layers of thematic maps, to determine the best possible route. The composite map indicates in a dashed line the corridor which can be utilized considering the existing features in each map and the corresponding restrictions.

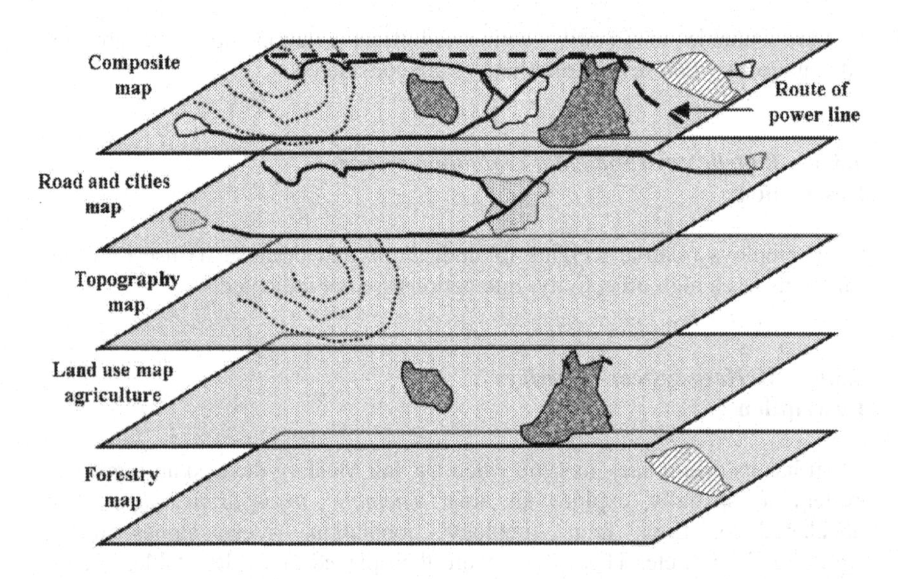

Figure 4.2. **Superimposing of thematic maps**

Thematic maps can be shaded to indicate some type of concentration for a certain activity.

This system is called *overlays* and obviously it has its physical limitations regarding the number of thematic maps and the clarity of the results. There have been many different applications of this model, one of the first being the selection of highway corridors. Once this map is overlaid on a geographical map (a receptor map), showing cities, roads, monuments, etc., it can be clearly seen which areas the *contamination affects, its extension*, and *related concentration with geographical distances.* It is something similar to the meteorological maps shown in TV news, showing which areas of a country are being affected for some meteorological phenomena.

The physical limitation in the application of this method is that no more than 10 overlays can be used. Today overlays are combined with Geographic Information Systems (GIS) using the same principle, but obviously, with an enormous amount of information that was not possible to obtain in the early days. See section 5.1 on GIS for more information about this system and an example of application.

Conclusion

This tool is objective, excellent for the analysis of *composite spatial interactions* and produces a *good synthesis.* For instance it is possible to prepare an impact map of some effect such as the diffusion of gases produced by large waste incinerators. This effect is obtained using *mathematical models* which spatially show the concentration of the pollutant. *Superimposing the concentration curves over a geographical area (receptor) it can be seen where the concentration is and its intensity.*

Uncertainties

None of the above discussed tools, except perhaps Battelle, contemplates uncertainties. Unfortunately uncertainties are a big issue in alternatives selection. There are uncertainties in many aspects like:

- *Type of environmental changes* that could be happening in the future due to natural causes, such as landslides, earthquakes, flooding, etc.

- It is impossible to know for sure what consequences will be produced in the future due to *human causes.*

- One has the *inability to reliably predict* the emergence and corresponding evaluation of the consequences of social-economic problems.

Uncertainty is closely related with risk analysis (see section 5.10), because many times it is not possible, because of uncertainties, to have a mean value and a standard deviation to work with. Probably the best technique is to use *sensitivity analysis* where variables are changed and results observed, so, in certain circumstances it is possible to ascertain the potential outcome.

4.3.7 The Delphi method

Description

It was developed after WWII by Olaf Helmer and Norman Dalkey of the RAND Corporation. It is essentially an iterative communication system that works with *groups of experts, spatially located,* in order to get their *independent* input about certain issues and responding to a coordinator. A *questionnaire* is prepared about some specific subject and then sent by mail or email to this panel of experts whose *members are not known to each other.* The experts are allotted some time to answer the questions posed, make their evaluations and comments and send them back to the coordinator. Once the responses of one round are received they are analyzed, aggregated, and then returned to the experts. The experts work with this information and refine their estimates.

The system continues until the responses from the panel of experts are getting close to each other, or in other words when the responses satisfy the entire group. Finally an average value is adopted. The system has proved very useful and more reliable and accurate than other conventional methods of forecasting, which is understandable considering that the experts in different fields of expertise are distilling years of experience and knowledge, and are *usually in a better position to forecast from facts rather than with theoretical methods.*

The usual forecasting mathematical methods are models that try to represent the real world as accurately as possible, which is not very often possible, and for that reason, and especially because they do not work with experience or subjectivity, their results are considered less reliable than

Delphi's. Experts on the other hand can do the same but also *adding a subjective input* which is not easy to consider with analytical methods.

Besides, they introduce their individual experience, which is impossible to take into account in mathematical models. The main characteristic of the Delphi method is the *anonymity of the panel of experts*, so, there is no influence from one to another, which is a definite advantage over methods such as *public meetings* where people can be influenced by others, especially if there is a leader. However, the Delphi method also has its drawbacks.

Related with the analysis for EIA, one of the problems could be that experts based in different locations could not know very much about the effects that a certain project would produce in a particular location. Another disadvantage of the Delphi method is that experts, because of their backgrounds, might not adequately represent or understand social issues, while the opposite can also be true in the sense that some experts could understand social issues but not be so knowledgeable about more technical aspects.

On the other hand open or face-to-face meetings have the advantage of *letting people consider,listening to other experts, about aspects that they have not thought about.* An additional advantage of face-to-face meetings is that they can be attended by *people living in the area who are going to be affected by the project.* They will be the direct receptors of the project's impacts, and obviously are in a position of discussing and having their opinions known about subjects that they are very familiar with. This first-hand knowledge of the situation is an input than can be complemented by the input from City Hall staff, with more knowledge about technical issues and difficulties. This concerted action will produce a result that is naturally a more complete assessment of every situation.

For instance, the construction of a Light Rail Transit System (LRT) in a city can cause a populated area to be cut in two, without easy communication between both parts. This is a fact that could have been overlooked by the analysts. As mentioned, questions are asked via a questionnaire and underlying all of these comments is the fact that it is extremely important the way the questionnaire is structured. If it is not clear or unambiguous the responses will answer questions that could have been not very well understood, or could produce different answers, which will make their comparison difficult. In other words the questionnaire should be structured with very clear and direct queries and without room for different interpretations of the same subject.

As was mentioned above, the Delphi method is an iterative procedure. Iteration, a very common practice in mathematics, involves the repetition or reiteration of a series of rounds or steps, with the characteristic that the response from round 2 is based on the answers provided by round 1. Response from round 3 is based from results from round 2, and so on. In this author's opinion the iterative nature of the Delphi method conspires against its reliability and accuracy, because in the successive rounds the experts receive information, that albeit in anonymity, can influence their decisions. Naturally, either the Delphi method or the open meetings method can generate answers that are perhaps not replicable, so an attempt to repeat a procedure with another panel or with other people will probably deliver a different result. The successive steps of the Delphi method can be summarized as follows:

1. State the project and define it as clearly as possibly.
2. Analyze the potential impacts of the project in different areas.
3. Select a panel of experts knowledgeable in the areas.
4. Prepare a questionnaire about the subjects which are important for the project, establish a date for the answers and send the questionnaires to the panel experts.
5. Once the responses have been received, analyze them looking for a reasonable coincidence of answers.
6. If a reasonable agreement has not been reached, prepare a document with the results from the prior round and feedback the information back to the panel experts.
7. Steps 4 to 6 are repeated until a satisfactory agreement has been obtained.

Conclusion

It is believed that even considering its drawbacks the Delphi method is an excellent tool to get *authorized information,* because the data comes from experts in each field, with time to think about an issue, consult bibliographic material, and make a written statement about their answers. Besides, *it is probably cheaper* than other methods but can take more time.

4.3.8 Dose-response functions

Description

The brief description of this subject only aims to give an idea of these tools and to make their existence known. It is beyond the scope of this book to enter into details about the production of these functions which are indeed a complex and a specialized issue and usually impact-specific. For the interested practitioner there are many publications on this subject where he/she can look for guidance. These are mathematical relationships often linked with risk assessment, and usually they relate the *concentration* of a chemical product with the *response* it is supposed to produce in a human being, animal, plants and even minerals.

For instance it is possible to analyze the effect of a certain concentration of SO_2 and NOx on some monuments, such as the Taj Mahal in India. Because this piece of art is made out of marble, the gases from automobiles form a chemical compound, H_2SO_4 which eats away the marble. An example of a dose-response function is shown in Figure 4.3.

The same applies to humans when one analyses for instance the effect that particulate in the air from a coal-fired power plant has on a nearby populated area. In this case a determination is made about the *response* of the human body to a certain *concentration* of the pollutant, and in principle it would appear that this relationship is linear, in that doubling the concentration doubles the damage. Generally it is not so, for the relationship between concentration and response is usually a curve, as in Figure 4.3.

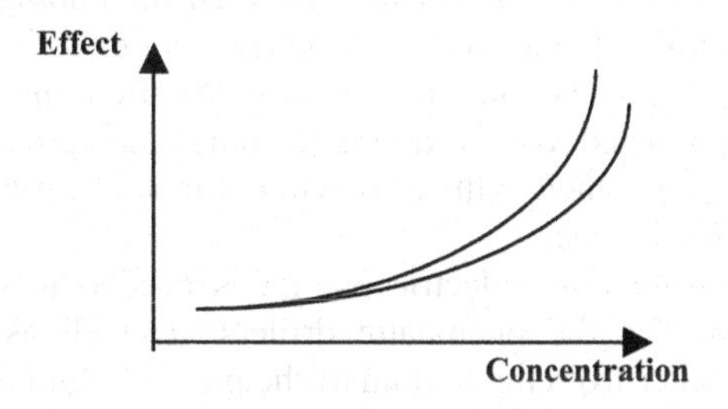

Figure 4.3. **Dose-response function**

The determination of this dose-response function is not easy but can be measured using lab animals. One expects that the curve thus designed can be applied to human beings, and it is, *but it does not mean that there is a certainty that the response in human beings is the same as in lab animals.* See in Appendix, section "Cited references on the Internet by industries", AREA: Software, the CALINEA4 and the CalTOX models.

4.3.9 Stepped or chained matrices

Description

Matrices (see Appendix section A.1.3) can be used to *show a cause and effect relationship*, because they can be combined in many matrices in a natural sequence. These matrices are called *stepped or chained matrices.* Assume for instance a mining and metallurgical operation. The project produces the following *actions*:

- Excavation;
- Transportation of ore to grinders;
- Crushing of ore in jaw crushers;
- Chemical and mechanical floatation process (see Glossary);
- Production of tailings (waste liquor from the chemical process, highly contaminated with very noxious chemicals, cyanides included);
- Refining operation (once obtained the metal it is upgraded to a commercial grade);

These actions provoke different *effects* such us:

- *Ground disturbance*, for instance in open pit mining the ore is extracted without the use of tunnels or galleries. A huge excavation finds the ore and it is transported through trucks following a twisted road bordering the hole in the ground;
- *Rock dumping*, which is the rock without metal bearing, and with no commercial value;
- *Energy consumption:* Electric energy is needed to operate the compressors for the pneumatic drillers, as well as pumps to discharge water from the bottom of the pit, and lighting, since the operation works around the clock, etc.;
- *Dust:* Large quantities of dust might be generated especially by trucks;
- *Noise generation:*Considerable noise is produced by compressors, the drilling equipment, the ore crushers, the conveyor belts and the trucks;
- *Fuel comsumption:* Large quantities of fuel are consumed daily for trucks and ancillary equipment;
- *Land use:* There is usually a profound modification of the land, because the excavation of the pit, the construction of roads, the construction of the metallurgical plant, etc;
- *Hazards:* This is a risky operation and there are many hazards such as:
    - -- Personnel accidents are frequent;

-- Surface sources of water get contaminated with dust.

In turn, each of these effects impact on workers and on the environment which are the *receptors* (man, soil, water, air, wildlife, vegetation, etc.).

The *consequences* of these impacts in these receptors are:

- Tailings are stored in what is called a *tailings pond* which can leak and contaminate aquifers;
- Tailings produce the death of many birds when drinking this very contaminated liquid;
- Dust produces respiratory problems.

Sequentially, these consequences produce birds migration and genetic changes, danger for a nearby town because of aquifer contamination, etc.

All of the above mentioned actions, effects, consequences and penalties, can be analyzed using a stepped or chained matrix, where *the output of one is the input of another.* It is also possible to place numerical values on each intersection in accordance with a certain scale, say from 1 to 10, and from expert opinion. This way the cumulative chain of events of a *single action* can be evaluated. Figure 4.4 illustrates the use of such a matrix.
Taking an action, for example a *chemical process* and considering *just one* effect such as *hazards,* a degree of *importance* can be expressed by a cardinal number between 1 and 10 (in this case 7). In turn, hazards can be related with different receptors, as mentioned above, with importance expressed by numbers in the hazards column. If for instance we only consider *water*, the chemical process contaminates large volumes, and an importance value of 5 is assigned; in turn, this highly contaminated water is piped into a *tailings pond*, but because its lining, its danger is low (importance 2), however, superficial water here cause the *death of birds* when drink it (importance 7), and also, because potential failures in the lining (importance 1) it can *contaminate the aquifer* which serves a nearby town.

4.3.10 Comparison of techniques

It is not considered that one technique or procedure is better than another one. Each of them have advantages and disadvantages, and many complement each other, although some of them are more suited than others for a particular application. Table 4.5 compares the different techniques and gives information about their characteristics.

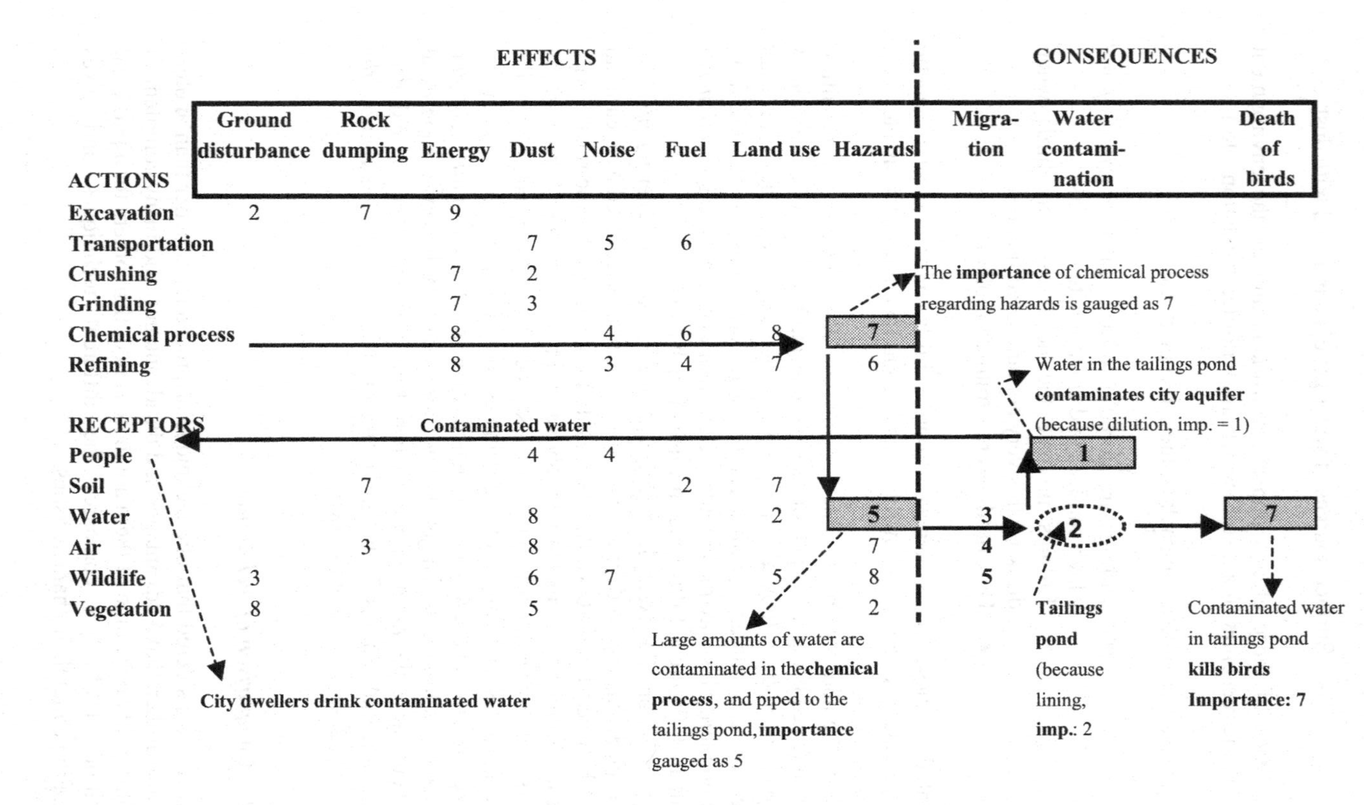

Figure 4.4. **Mining and metallurgical operation**
Determination of effects and consequences through a stepped matrix

Table 4.5 **Comparison of techniques**

Techniques	Advantages	Disadvantages
Checklists	- Simple to understand by everybody	- Do not show indirect or cumulative effects
Networks	- Provides an effective and visual way to show cause-and-effect relationships	- The system could become very complicated to examine but can be improved converting it in a matrix
Leopold's matrix	- Forces to a systematic revision of the project and its impacts	- Too many values to be considered for appraisal. - It does not consider time-dependent effects
Overlays	- Very good spatial information. - Good for interaction between impacts	- Expensive - Cannot identify secondary and tertiary impacts
Delphi	- Expert opinion on different issues	- It is believed that expert opinion can be influenced by results from former rounds
Dose-response functions	- Establish a direct relationship between effects and responses to them	- Need experimentation to be developed It is not certain than tests in animals obtain the same response in humans - Impact specific
Stepped or chained matrices	- Provide a very clear representation of a cause and effect relationship in a natural sequence	- They can become a little complicated when there are many elements involved

Internet references for Chapter 4

CONCEPT: Check lists
Title: *Assessment of Environmental Impacts for Irrigation Projects. A Decision Support System*
Authors: Abu-Zeid, Khalem M, Bayoumi, Mohammed N & Wagdy Ahmad.
Presents a listing of typical impacts of dams and reservoirs, together with a good analysis of impacts for irrigation projects.
http://www.dams.org/kbase/submissions/showsub.php?rec=opt081

CONCEPT: The Delphi Method
Title: *The Delphi method - Definition and historical background*
Provides good information about the history and development of this method. and details of the basics of the Delphi method. This paper is an excellent

introduction to this tool for collecting and evaluating information, which is a crucial point in EIA.
http://www.iit.edu/~it/delphi.html

CONCEPT : The Delphi Method
Title: *Computer based Delphi processes*
This is a very extensive paper incorporating:

- Asynchronous interaction;
- Anonymity;
- Moderation and facilitation;
- Structure;
- The trend model;
- Analysis;
- Scaling methods;
- Structural modeling;
- Delphi, Expert Systems, GDSS, and Collaborative Systems;
- The deductive and inductive levels of disagreements;
- Goal and value disagreement.

It is an excellent introduction to this methodology.
http://eies.njit.edu/~turoff/Papers/delphi3.html

AREA: Ecosystems
Extensive publication that provides Dollar-based Ecosystem Valuation Methods. The paper describes methods linked with ecosystems as follows:

- Market price methods;
- Productivity methods;
- Hedonic price method;
- Travel cost method;
- Damage cost avoided, replacement cost and substitute cost methods;
- Contingent valuation method;
- Contingent choice method;
- Benefit transfer method.

For each one offers a detail explanation, steps to implementation and provides several examples of case studies.
http://www.cbl.umces.edu/~dkingweb/contingent_valuation.htm#summary

CONCEPT: Batelle Environmental System
Title: *Detailed assessment of environmental impact - Description of a forestry project*
http://www.fao.org/docrep/t0550e/t0550e06.htm

CHAPTER 5 - TECHNIQUES FOR ENVIRONMENTAL APPRAISAL

5.1 Geographic Information System (GIS) Description

What is a GIS?
It is a specialized software with the capability to deal with spatial information.

5.1.1 Fundamentals

One of the problems with the different methodologies used to appraise projects is that *they are not spatially related.* This means that they analyze an impact and put values on its effects but *do not geographically relate the impact with the area it affects.* EIA needs a large amount of information for the analysts to make their decisions, and certainly most of this information is space or geographically dependent, such as the diffusion and concentration of pollutants, the effects of erosion, the existence of a forest and number of birds in it, etc.

Geographical Information Systems (GIS) is on the other hand admirably suited to manage information when it is geographical referenced (georeferenced), for it can:

1. Store data on different issues, for instance land, flora, rivers, mountains, etc;
2. Amalgamate diverse type of data;
3. Display this data in a screen, with the capability of identifying it regarding its characteristics.

Objects represented in maps are called *features,* and a spatial relationship shows associations between feature locations. GIS stores all the information about map features in a GIS database and then links them.

Information is stored as attributes in the form of tables (see Table 5.1). For instance, for a river, its attributes are: length, width, flow, dams, bridges, etc. Features can be displayed on a map based on any attribute. For instance traffic can be shown for the road, and between certain locations, and, at the same time, width of the road and number of lanes or posted speed, etc.

Table 5.1 **Linking attributes and features**

Attributes:	Length	Width	Flow	Bridges	Surface type
Features:					
White River	x	x	x	x	
Road # 412	x	x			x

GIS can exhibit a set of features and link them together in what is called a composite map, that is a set of different features sharing some attributes. An example can be the linking of a national park with attractions, roads, wildlife habitat, etc., in it or nearby. Thematic maps can be displayed for:

- Agricultural land;
- Rivers;
- Populations;
- Slopes;
- Parks;
- Geological features;
- Wetlands;
- Bird sanctuary;
- Protected area;
- Etc.

Once a group of features is selected a series of operations can be performed, such as getting *cumulative data*, which is very important for EIA. Most important for the analyst is to determine which features share the same geographical space, as was discussed in overlays. Utilizing different thematic maps GIS uses the location data stored in its database for each feature, regardless of its thematic nature, and finds the *intersection of the different features*. Understandably there are several subsystems that are interrelated into GIS.

As an example suppose that there is a need to determine a location for the installation of hazardous waste incinerators. These locations have to comply with a series of prerequisites such as being away from populated areas, but not too far away from them either, close to commercial roads and highways, etc., and that can be found using GIS. At the same time an overlay map could show the direction of prevailing winds which can transport particulate. Information might also be colored to indicate diverse characteristics, like the different composition of soils in an area.

Satellite images can be incorporated into the system showing features such as cultivated areas and identifying different or similar types of crops. This is important in *spatial correlation* to avoid the problem of autocorrelation. For routing determination, GIS is extremely useful. Suppose that a central government wants to build a rail link between two industrial cities over a distance of 73 km. There is a potential corridor which offers different alternatives with variations in distances and costs. Different features exist between both cities such as roads, towns, rivers, channels, heritage sites, different terrain slopes, a bird sanctuary, etc.

Using GIS, all of these features can be assembled together, as well as the different alternatives for the intended track. As this information is georeferenced, it is possible to superimpose data and get a composite drawing of an area. The resulting maps, for each section of the terrain, can give very valuable information and help the taking of a decision. GIS systems are extensively used for:

- Inventorying land use, as for instance finding out the area of cultivated land;
- Determination of type of crops in each area;
- Identification of boundaries between different land uses;
- Verification of population density;
- Production of cadastral drawings;
- Inventorying utilities buried in a street;
- Etc.

These remarkable characteristics are extremely useful for Environmental Impact Assessment, for several reasons. One of them is that it is possible to get a *baseline of an area* before the effects or impacts of a project take place, and it is also feasible to simulate the effects of some impacts and see how the area is influenced. Information can be gathered through direct observation or by *remote sensing,* including aerial or satellite observations and maps. In cities GISs are extensively used to store information about buried utilities, size of each lot, taxes paid, improvements made in properties, etc.

A word of caution

In large projects involving perhaps thousands of kilometers, typically for gas and oil pipelines, aqueducts, transmission lines, etc., care has to be taken in the kind of maps used. Maps are two-dimensional representations or *projections of features* from our quasi-spherical world. Because of its curvature, it is impossible to "flatten" it, and different types of projection systems have been devised in order to translate the geographical coordinates of a feature on the sphere, into a sheet of paper. Most important projections are:

- Mercator
- Robinson
- Equidistant cylindrical
- Sinusoidal

Each one has its own characteristics but none of them can represent Earth as it is. Some projections distort more in the Poles and others more in the Equator, however, the Mercator projection is probably the most used. GIS can work with any of these projections or even others but the most important aspect to remember is that *data in all the regions have to be based on the same type of projection.* This requisite is understandable when one considers the different type of distortions that each system produces, so merging two different projection systems will produce breaks and misalignments.

5.1.2 Case study --- Road construction near archaeological ruins

Background information

To illustrate the use of GIS imagine the following example. In the Yucatán Peninsula in Mexico, there are many ruins of great archaeological value all over this territory. There is a project to build a road between the resort town of Playa del Carmen on the east coast, and the State capital city of Mérida. The most direct layout passes very close to some of the most important Mayan ruins such as Cobá. The proposed route will start at Playa del Carmen and will intersect highway 180 to Mérida in the town of Chemax. The total extension of this new route is about 73 km. To start working, baseline data was collected along this route within a corridor 20 km wide. This information was then introduced in a GIS system such as Arc Info® and ArcView®, and thematically grouped as follows:

- Geological formation;
- Stability;
- Nearby villages;
- Agricultural land;
- Distance to Tobá ruins,
- Kind of vegetation in the area;
- Existence of desert zones;
- Slopes;

- Volume of vehicles per hour that will use the route;
- Rivers;
- Prevailing winds;
- Normal atmospheric conditions.

The whole length of the route was then divided in three areas of about 25 km. each in extension. For each area, thematic maps were developed and then electronically superimposed. There were no major problems except the crossing of some agricultural land. The route passes at 1.3 km from the Tobá ruins, therefore the physical construction of the route was not affecting them.

Considering the forecasted traffic it was possible to determine the generation of pollutants such as SO_2, and others which produce acid rain; a diffusion model was built using known techniques, and a map of this diffusion was prepared and introduced into the GIS database. When this map was layered over the baseline map of the Tobá ruins, it was found that there was danger posed by the pollutant concentration over this *receptor*.

Therefore, the trace or outline of the route was changed, a new diffusion model and a new diffusion map were then created and superimposed to the ruins. Subsequently, it was found that under the most severe atmospheric circumstances there would be only a very small amount of concentration over the ruins, and below the corresponding threshold.

Besides, geological maps also showed a weak soil condition, created by the existence of caverns, that motivated changing the route outline again, for this highway will be used for heavy traffic as well as cars. In order to compensate for the increased production of CO_2 by traffic, it was decided to plant thousands of trees along the route, which can function as CO_2 sinks (see Glossary).

Conclusion

GIS is the only technique that graphically shows the *superimposing,* and then the *cumulative effects of many different impacts*, and in many occasions can be also used to select alternatives especially in cases involving long distances such as in power lines, aqueducts, roads, etc. Besides, it also georeferences them so its value doubles, and another advantage is that in some circumstances it can identify areas that comply with established thresholds. However it does not have the capability to indicate secondary and tertiary effects.

5.2 Contingent Valuation (CV)

Description

There are circumstances in which the values of some environmental assets such as clean water, land, clean air, etc, as well as benefits we receive from ecosystems such as watching wildlife in tropical forests, ice fields, etc., are difficult to evaluate in economic terms. One way to solve this difficult issue is to assign a dollar amount to these benefits, *not in the sense of allocating to them a market value,* but instead assessing how much they mean for people. For instance how much is somebody *willing to pay* for the privilege of walking in a Costa Rican forest?

Contingent evaluation does precisely this, and it is called "contingent" because people are asked for *their willingness to pay, contingent upon something being obtained in return.* Notice that the aesthetical pleasure of bird watching is something that has no marked value, cannot be purchased or sold, or used, and for that reason has no "use value". In this context we can buy and use oil, gas, water, land, cleaner air (for instance by moving to a part of town with cleaner air than where we live), etc., but we cannot buy the sight of majestic icebergs in the Antarctica. How is it possible to measure people's willingness to pay for a contingent good? By surveys.

But this is not easy. Surveys can be made through interviews, that is face to face, by mail or email, by telephone, etc, but in any case the questions have to be thoroughly clear, and of course, providing enough background information and data and specifying exactly what we are trying to find. Remember that the surveyed persons have to understand what the survey is about. The number of people consulted can be determined using statistical sampling where a level of confidence as well as precision values are established. There are many manuals about the determination of the size of this sample, so the interested reader can get information from numerous publications or from firms specializing in surveys.

The result of the survey will be a statistical value for each question posed, and in essence it reduces to a comparison between how much people are willing to pay and the costs involved. An interesting case was the damage evaluation made by the U.S. National Oceanic and Atmospheric Administration (NOAA) , involving the accident of the Exxon Tanker "Exxon Valdez" which spilt oil off the coast of Alaska. Another case was the evaluation regarding contamination by industrial wastes along the coast of Southern California that contained DDT (Dichlorodiphenyltrichloroethane) insecticide, and PCB (Polychlorinated Byphenyls') from electrical and industrial applications.

In both cases the Contingent Valuation Method was employed, and as a consequence of these actions the NOOA published a "Natural Resource Damage Assessment (NRDA)" rule. For more information please access: http://www.sasayama.or.jp/english/opinion_e/S_E_16.htm#honbun

5.2.1 Case study --- Visitors in an environmentally sensitive site

Background information

A system of caverns had been discovered 80 years ago, and for the last 35 years visitors were allowed to enjoy it for free, only paying a nominal fee for a tour guide, with the opportunity to walk on selected trails and to observe huge stalagmites, stalactites, underground rivers and lakes, some rare species of insects, etc. Unfortunately the increasing number of visitors was causing rapid deterioration in the caves, because human bodies introduce humidity into this ecosystem, and touching the stalagmites and stalactites degrades their surfaces by the natural body oils deposited. Besides, the normally constant temperature in the caves began to fluctuate with the presence of visitors' body heat.

So, a project was undertaken to protect the site, by providing more security, establishing a mandatory circuit in the caves using electric cars, preparing the people through a prior lecture before the visit, *and mainly restricting admitttance to only a certain number of people per day.* The estimated cost of this project was about US$ 480,000 per year. A survey was conducted using travel agencies who sent visitors to the caves. It was requested that these organizations ask interested people how they would feel about contributing with a fee of US$ 7 per visitor, to be paid now *to maintain the caves for generations to come.*

Also a survey was conducted by mail, asking the same question of people who had previously visited the caves and were registered as visitors. The whole survey took 18 months to be completed, and it showed that 82 per cent of people consulted agreed that it was worth paying for something that their descendents could enjoy.

As mentioned, one of the conditions for the new procedure was to admit only a maximum number of visitors per day and during the eight hours of operation. There is no lack of tourists, and a restriction was to be enforced to permit no more than 200 people per day, so the maximum amount of visitors per year was 200 x 365 = 73,000 persons. Since the entrance fee was fixed in US$ 7 per person, the annual income was projected to be 7 x 73,000 = US$

511,000, enough to cover the costs. Therefore and because of the willingness of people to pay, the caves remain open.

Conclusion

This important tool deals with *intangibles*, that is, when the value of an asset is not known --- or it is *not possible to put a market value* on an asset--- and, at the same time, when it is impossible *to place a value on the benefits it produces.* This technique works by assigning a value to these benefits, albeit not a market one, but determining the people's willingness to pay to enjoy what this asset offers. In other words, people are willing to pay some amount of money that is contingent upon receiving something else in return, even if this return is non-tangible.

5.3 Cost-Benefit Analysis (CBA)

Description

The implementation of projects, programmes and policies has an impact on society. Cost-Benefit analysis is a collection of methods designed to estimate the *social consequences produced by this implementation.* Given a defined project it is possible to use CBA to appraise the expected impacts it will produce measured in economic terms. This concept is similar to the CBA performed to evaluate a certain commercial project, in that it balances gains and losses that take place at different moments, and converts them to a single value at the present time.

However, a commercial project is expected to yield an appropriate return to the firm which promoted it. In other words a commercial CBA makes a *comparison of benefits* and *costs,* both expressed in monetary terms, which have to be discounted over time. In the commercial context, the benefits will be the net dollar gains expected and the costs will be the expenditures that various choices of action will incur.

CBA for commercial appraisal and CBA for environmental appraisal are similar but some concepts vary in their point of view. In the EIA context, benefits are derived from the *users' willingness to pay* (WTP) for them. Likewise, costs are based on the *users' willingness to accept* (WTA) a monetary compensation.

There is also a difference regarding the valuation of costs and benefits. In a commercial project, benefits and costs are taken at *market prices,* but in considering costs and benefits for a project involving environmental and social aspects, market prices are usually unsatisfactory. The reason for this is that normally *market prices do not represent the interplay between offer and demand as in a free market economy*, because of distortions. Examples of these distortions are:

1. In many countries the government establishes a top price for fuels. Then the fuel price will not reflect what the economy as a whole is paying for it, but an arbitrary value.

2. A monopoly can set prices for goods that do not represent the actual values of their inputs.

3. Some countries do not have a free market for foreign currency, so their values do not reflect the laws of the market.

For this reason, if CBA is used to measure the profitability of a project, it is necessary to utilize the real cost and prices, which are expressed as *shadow prices* (see Appendix, section A.7), which is a very well known concept in economics. The Net Present Value method (NPV) can be employed to compute this profitability, but in the case under examination the benefits and costs should not be discounted at the prevailing interest rate of the market but using the *cost of opportunity of capital*, (see Appendix, section A.6), which is the interest that society would get if those funds were used in other ventures.

As expected, the estimate of these shadow values is not an easy task, and their calculation is beyond the scope of this book. A main problem is estimating the gains and losses and determining *who* are the people who will get a benefit or will incur a loss; but probably most important is the *calculation of a monetary value corresponding to the benefits and losses without market value.* In order to understand the literature and concepts of Cost-Benefit analysis it is convenient to have some expression defined. The definitions given here are very elementary and their only purpose is to make the reader aware of their general meaning. Needless to say, to fully understand these concepts of Environmental Economics the reader might want to consult books on this matter, which should be available in any library.

Price:
The amount that is paid for something.
Value:
The amount of satisfaction derived from the use of something.
Consumers' surplus:
The excess on the price, paid for something.

Market value:
Pertains to tangibles (houses, cars, airline tickets, etc) that have a value in the market and that, as a consequence, are subject to the laws of supply and demand.
No market value:
Pertains to intangibles such as the beauty of a beach, the view from the top of a mountain (also noise, pollution, etc).
Externalities:
The impact produced by individual action *without the intervention* of the market. Assume that you are standing at the intersection of two avenues and waiting for the traffic light to allow you to cross the street.
Suddenly, a heavy truck passes in front of where you are standing and accelerates so as not to be caught by the red light and in so doing hits you with a cloud of gases and particles. So you (the receptor), *are receiving the impact* produced by the *action of the truck driver* (the stressor) which cannot be measured, i.e., *there is no market intervention.*
This is called *an externality.* Typical externalities are air contamination, noise, vibration, temperature increase, dust, etc.
OLS: Ordinary Least Squares: Statistical technique to find the relationship between two variables. It is mostly known as Regression Analysis (see Appendix, section A.1.1). The solving of these systems leads to obtaining a linear regression equation with coefficients representing the shadow prices.
Shadow prices: The change in the price or objective of something due to the unitary change of one of its attributes, holding constant the balance of the attributes (see Appendix, section A.7).

CBA is a difficult issue and diverse methods have been devised to find these values, most of which use indirect ways. These methods are known as:

- Contingent valuation (see section 5.2) where people are asked about their *willingness to pay* (WTP) for a benefit and the *willingness to accept* (WTA) compensation for a loss.
- Contingent ranking: In this method people are asked to establish a ranking among several options.
- Stated preferences: A value for a place, say a national park, is established based on peoples' preferences.
- Travel cost: It is assumed that the value of a site can be obtained considering how far away it is and how much it would cost for people to reach it. An example could the Grand Canyon, in USA, which has such a high value that people come from around the world to visit it, just for sightseeing, or for other recreational activities.

- Hedonic pricing.

5.3.1 Hedonic pricing

Probably the most important and most widely used method. In essence it makes an estimate of the *change in price* of a *market good,* such as a house, when *changes occur in some of the attributes* that characterize such a market good. For instance, it might need to evaluate how much the price of a house *would decrease (*i.e. *a change of its market value)* due to a change of some attribute *(*for instance *the disturbance of the quietness of the site)*, caused by the construction of a new highway nearby.

The Dictionary defines the word *hedonic* as "something characterized by pleasure". From the economic point of view it is related to the pleasure given by some environmental assets such as viewing beautiful scenery, watching animals in their natural habitat, scuba diving in a coral reef, fishing in a forest creek, etc.

Hedonic pricing is a relatively modern concept designed to put some values to these environmental assets. It originated in the seventies with the seminal work by two researchers, Zvi Griliches (1967) and Sherwin Rosen (1974), (see Bibliography -- Griliches Z. and Rosen S.). Hedonic prices are most often related to Real Estate or labor markets, and one of the reasons for this preference being the large database existent for both contexts. For instance, there is abundant recorded information about the effects that some externalities such as noise, air contamination, traffic, etc., produce on people and on the value of their houses, and there are records comprising many years of sales prices.

From the labor point of view there is also copious data on the subject of how income is related to economic conditions, new technical developments, demand, etc. The fundamental concept behind hedonic prices is that the value of a good does not depend only on its price but is also a function of other characteristics; that is, its *value* is a consequence of many factors that are bundled together in some selected *package.* This package includes characteristics such as:

- *Physical attributes* of the house (size, number of bedrooms, garden, existence of a driveway, garages, etc.). These physical attributes are even broken down by their individual characteristics such as size of bedrooms, garden dimension, number of cars that can be parked in the garage, etc.

- *Accessibility* and *location* attributes such as distance to schools, shopping centers, public transportation, parks, etc.

- *Amenities* like view, temperature, pleasant environment, privacy and quietness, etc., and *disamenities* such as heavy traffic, unpleasant neighborhood, crime rate, etc.

All of these characteristics have a certain weight and they are not considered alone, but *bundled together* to determine the *value* of the house. Considering the above, it is then possible to determine the influence or weight that a disamenity has on the house value.

Suppose for instance that a highway is to be constructed very near a given property. The impacts are twofold: construction impacts by generating dust, noise, lack of privacy, heavy traffic, etc., and the operation phase impacts with noise, garbage, traffic, accidents, traffic jams, etc.

Most definitely the value of the house will decrease because of this construction; in other words, there will be *impacts that will reduce its worth.* If the extent of noise, loss of privacy, and other such characteristics is known, then it is possible to calculate the *decrease in price.* We can reach a result that tell us that the decrease in price because of the highway construction produced a decrease of say 12 per cent.

The problem, of course, is how to compute the weight or contribution of each characteristic that affects the value of the house. Hedonic prices correspond in reality to the *implicit prices of the attributes*, and these prices can be estimated by applying Ordinary Least Squares or OLS for short, which is in essence a *linear regression* (see Appendix, section A.1.1) of the total price of the house considering the prices of all the characteristics.

A linear regression equation has a constant or independent term, and a factor affecting each characteristic. The constant term corresponds to the intersection of the regression line with the price axis, and the factors or coefficients represent the shadow prices, that is the change in the price of the house for one unit of change in the corresponding characteristic.

Hedonic prices have been especially used in transportation projects, because as expected, they produce many externalities alongside their length.

5.3.2 Case study -- Sewage conversion in a tourist resort

Background information

This example considers the case of a resort in South America, on the Andes Range. It receives visitors the year round from all over the world, who stay in a village and in camps. This village, populated by about 1,700 permanent inhabitants is the hub of the area, and is located in a national park, on the shore of a lake. Its only revenue comes from tourism. The area is a first class recreational center, and offers opportunities for fishing, sightseeing, water sports, trekking, climbing, skiing, hunting, whitewater boating (in a river feeding the lake), etc. The whole zone lies on a hydro-basin. There is abundant and varied wildlife and a large fresh water fish population, especially trout.

Water for domestic purposes is plentiful in the area and there are no industrial plants, except some home industries that produce a chocolate which is famous in the country. The village has a sewage network but other than a primary filtering to remove solids, the waste is discharged into the lake without further treatment. With the increase of tourism this discharge creates a serious problem not only for human health reasons, but also for the fish population, and because it causes turbidity in the water tourists are starting to complain about the lack of transparency of the lake water, the absence of fish in certain areas, and the smell.

Even when drinking water is extracted from the lake from a distant location and then purified, there are concerns about its potential contamination. There is a national plan to convert the area into an ecological park in order to preserve its beauty for future generations. This measure will impose severe limitations on new housing developments as well as very stringent restrictions on the treatment of waste water. Therefore, members of the population are consulted about their preferences regarding how to stop deterioration. The government proposes, and it was accepted by nearly eighty five per cent of the population, the construction of a small water treatment plant, with a capacity to treat waste for up to 15,000 people.

The government does not have the money to finance this scheme, and as a consequence is asking the population to bear its expense, that is, their *willingness to accept costs.* A study has been made regarding the potential benefits that the village will receive, translated into attracting more visitors because of improved environmental conditions. Therefore the citizens have been consulted about their *willingness to accept benefits.*

The calculation is based on the following data:

Background information for the computation of benefits:

Average annual rate of increase of visitors:	1.6 %
Visitors baseline (year 1):	56,800 persons
Average spending per day:	US$ 73
Average number of stay per visitors:	4.8 days
Job creation:	16.42 visitors/employee.

These figures estimate how many local people are needed to attend to visitors' necessities, and include people employed in:

- Hotels and camping;
- Restaurants;
- Supermarkets;
- Shops;
- Entertainment industry;
- Excursions;
- Guides;
- Etc.

Background information for the computation of costs:

Total number of dwellings:	426
Dwellings growth:	3.8 % per year
Cost to build the new water treatment plant and new trunks:	US$ 10,521/household
Payback period:	10 years
Discount rate :	7.68 %

Regarding the last row, this is not the discount rate based on the opportunity cost (see Appendix, section A.6) because in this case the people are paying for the construction, not the government. As a consequence this is the *market rate.* It has been agreed with the people that the new system will begin operating in the first year. The project asks for a two year period for the homeowners to be connected to the new system.

The survey also showed that in the first year only 25 % of the dwellings will be connected, and the remaining 75 % in the second year. Obviously, new constructions will be automatically linked to the new water treatment plant.

Table 5.2 condenses this data for the calculation of benefits.

Table 5.2 **Computation of benefits**

Year / Concept	1	2	3	4	5	6	7	8	9	10
Total expenditures from visitors (in million of US$)	20	20	21	21	21	21	22	22	22	23
Jobs creation (in hundreds)	3,4	3,5	3,6	3,6	3,7	3,7	3,8	3,9	3,9	4
Economic benefits of job creation (in millions of US$)	7	7	8	8	8	8	8	8	8	9

Table 5.3 shows the computation of costs.

Table 5.3 **Computation of costs**

Year / Concept	1	2	3	4	5	6	7	8	9	10
Total number of dwellings	426	442	459	476	495	513	533	553	574	596
New dwellings		16	17	17	18	19	20	20	21	22
Average payment per dwelling for the Water Treatment Plant (x 00)	10,5	10,5	10,5	10,5	10,5	10,5	10,5	10,5	10,5	10,5
Total costs (in million of US$)	1.1	3.5	0.177	0.184	0.190	0.198	0.205	0.213	0.221	0.230

Table 5.4 shows the final computation, that is benefits minus costs.

Table 5.4 **Benefits minus costs**

Year / **Concept**	**1**	**2**	**3**	**4**	**5**	**6**	**7**	**8**	**9**	**10**
Benefits - Costs (in million of US$)	6.2	3.9	7.4	7.5	7.6	7.7	7.9	8	8.1	8.2

Net Present Value: $ 784,565

Obviously the proposed new water treatment plant will produce indirect benefits to the area, so the construction obtained the green light.

Conclusion

Cost-Benefits analysis is a very useful and much used tool which amalgamates two issues:

1. A *non-market value indication of accomplishment*, that in the proposed example is the enhancement of a tourist area;

2. A *market value estimate of costs involved*, represented by the expenses needed to execute certain infrastructure works.

5.4 Cost-Effectiveness Analysis (CEA)

Description

Many times the objective cannot be valued but can be defined. For instance an industrial plant dumping contaminated water into a stream. The objective can be defined as to have a cleaner stream after the discharge of waste water, but *no value can be placed on this objective.* CEA *seeks to meet the objective at the least cost.*

Generally it works by computing the *potential reduction of contamination* that can be achieved with different schemes (that is the gain in environmental benefits), and compares it with the *annualized costs* of the different alternatives that can be employed to obtain this. The alternative producing the least cost is then selected.

5.4.1 Case study --- Selection of exhaust filtering equipment

Background information

A car manufacturing company wishes to improve the emissions from its sets of Coppola Furnaces. These devices, used for smelting purposes for a certain alloy, spew contaminated emissions, one of the main components being particulates. A technical study showed that by installing some specialized filtering mechanism the contamination can be drastically reduced. The Clean Air Act encourages this type of improvement by facilitating loans with a low rate of interest. There are three classes of filters and cyclones that can be installed, identified as Type A, B and C. Table 5.5 describes the characteristics of each one as well as the arithmetic operations involved.

Explanation of Table 5.5:

First row:

This is the baseline, that is the condition the system is working at present without any improvement.

Second row:

Shows the different gains obtained with each class. Thus Type A reduces emissions by 50% , Type B by 44 % and Type C by 51%.

Third row:

It shows the gain in reduction as per the values of the second row. Obviously the most efficient is Type B.

Fourth row:

Assumes that the system works during three hundred days per annum.

Fifth row:

Is the gain in reduction of contaminants in 300 days.

Sixth row:

Details the price of each piece of equipment, Type C being the most expensive.

Seventh row:

Shows the promotional interest rate loans, which is the same for the three choices of equipment.

Eighth row:

Shows the annualized cost per piece of equipment computed using the annuity form described below.

Ninth row:

Compares the annualized cost with the annualized gain in reduction.

The system is simple. It compares the gain in contaminant reduction (which is obtained for each class of equipment from its manufacturer), with the total cost for each one that also includes its maintenance.

Table 5.5 **Calculation for different classes of exhaust filtering equipment**

	Type A	Type B	Type C	Units
Baseline emissions of particulates	0.150	0.150	0.150	mg / Nm^3 – 24 hr
Emission reduction of particulates	0.075	0.066	0.076	mg / Nm^3 – 24 hr
Gain in reduction	0.075	0.084	0.074	mg / Nm^3 – 24 hr
Days of operation in a year	300	300	300	days
Gain in reduction per year	22.5	25.2	22.2	mg / Nm^3 - year
Cost per class	282,000	289,400	290,000	$
Promotional rate on interest as per government plan for clean air	4	4	4	%
Annualized cost over three years	101,618	104,285	104,501	$
Cost-effectiveness	**4,516**	**4,138**	**4,707**	$ / mg-year

Since there is a promotional rate of interest for loans from the government, and considering three years for amortization, one can use the following formula, available in any financial literature, to calculate the annual capital recovery:

$$\text{Annuity} = PV\,[\,i\,(1+i)^n\,] \;/\; [(1+i)^n - 1\,]$$

where:

PV = Present value
i = Interest
n = Number of years
The expression: $[\,i\,(1+i)^n\,] \;/\; [(1+i)^n - 1\,]$ is the "recovery factor" that in this case is:

$$[\,0.04\,(1+0.04)^3\,] \;/\; [(1+0.04)^3 - 1\,] = 0.36035$$

Replacing for Type A:
Annuity = 282,000 x 0.36035 = $ 101,618

Replacing for Type B:

Annuity = 289.400 x 0.36035 = $ 104,285

Replacing for Type C:

Annuity = 290,000 x 0.36035 = $ 104,501

Now the comparison is made between costs and gains:

Cost-effectiveness (Type A) = 101,618 / 22.5 = 4,516 [$/mg]
Cost-effectiveness (Type B) = 104,285 / 25.2 = 4,138 [$/mg]
Cost-effectiveness (Type C) = 104,501 / 22.2 = 4,707 [$/mg]

From the cost-effectiveness point of view, Type B *is the most convenient, since it removes one mg of particle at the least cost.*

Conclusion

This is a simple but a very important tool for analysis because it allows the integration of two fundamental concepts: *Economic development* and *environmental objectives*. It combines a *market value for costs* with *non-market values for environmental gains obtained.* In many cases it can give an answer to the question: *How much are we willing to pay for a sustainable environment?*

5.5 Input-Output analysis (IO)

Description

Input-Output analysis refers to the use of a tool developed in 1936 by Wassily Leontief, in a work entitled *"The structure of American economy, 1919-1939"*, 2nd edition, Oxford University Press, New York, 1951. In essence the model uses a matrix or table format, with an equal number of rows and columns, each of them representing one sector of the economy of a country. It is also called the industrial interrelationship matrix because it shows goods transferred to one industrial sector from another industrial sector.

So, if in a certain row we have say the food sector, the intersection of this row with all the columns, indicates the purchases of the food sector from all the other sectors, including the food sector. Almost every country prepares IO Tables to represent its economy. The tool, which is in reality a matrix (see

Appendix, section A.1.3), uses linear algebra and can be readily solved, provided that we get the inter-industrial coefficients.

When the IO model is used to analyze the production of contaminants it is called Environmental Input-Output and identified as EIO. In this context the IO tables are combined with environmental information from the Life Cycle Analysis (see section 5.6). In the following case example the EIO matrix is used to compute the contamination generated for the automobile industry. A pioneer in this type of calculation has been the Carnegie Mellon University Green Design Initiative (see Internet references for Chapter 5, at the end of this chapter).

5.5.1 Case study --- Contamination produced by the automobile manufacturing industry

Background information

To illustrate the method suppose that one wants to compute the amount of pollution created by the manufacturing of automobiles in a country. Assuming that all plants in this country assemble 800,000 passenger cars per year, their manufacture (backward cumulative effects), releases contaminants which can be calculated with the IO technique. This computation considers ALL SECTORS of the economy involved in the construction of motor vehicles, such as for instance:

- Mining operations to extract minerals for producing steel, aluminum, and copper products that go into each car;
- Blast furnaces and steel mills (steel production for car body and foundry for engines);
- Chemicals and coke used in the blast furnaces;
- Generation of electric energy to operate all the equipment used in the progression (appropriately called the *supply chain);*
- Primary aluminum (to make several parts of a car, as well as engines);
- Railroads and related services (to transport minerals for blast furnaces, finished cars and parts, etc,);
- Trucking (for transporting finished vehicles and raw materials and parts);
- Motor vehicles (personal use of motor vehicles for this production, for instance cars utilized by personnel assembling the 800,000 cars);

- Air transportation (personnel and parts);
- Glass (to make windshields and windows);
- Automotive stamping (to create the body);
- Natural gas (for furnaces);
- Compressed gases (for welding);
- Chemicals (for paints);
- Fuels and lubricants;
- Nitrogen and phosphate fertilizers (for crops used to feed people making cars);
- Plastic materials and resins (for plastic parts and for gluing);
- Paper and paperboards (for packing parts and others);
- Wholesale trade;
- Etc.

So, in this incomplete listing representing the supply chain, the whole scope of activity sectors involved in the construction of a car can be seen. In the IO model each sector requires something from another economic area, which in turn needs a little bit from other sector, that in turn necessitates something from other area, and so on. For this reason the logistics of the entire process is called the *supply chain.* In this example, instead of detailing the needs for parts or products, the model will consider and determine the amount of pollution that the production, fabrication, construction or manufacture of each of these components creates.

This is a linear model, therefore the total generation of pollutants will be directly proportional to the number of cars manufactured, and the fundamental data to introduce is the *amount in dollars* needed for producing 800,000 cars. Assuming a production cost of $ 13,867 each per car, it makes a total of $11 billion. The following figures have been obtained using a procedure developed by Carnegie Mellon University Green Design Initiative (2003), Economic Input-Output Life Cycle Assessment (EIO-LCA) model, which is available from: http://www.eiolca.net/.

Entering with the $11 billion figure we obtain the information shown in Table 5.6 for *contaminants released to the atmosphere for this production.* The procedure to get these values is as follows:

BACKWARDS EFFECT

1. Enter the Carnegie Mellon University Green Design Initiative (2003) at their www.eiolca.net web site.

2. Browse under "Categories" for the area of interest. In this case it is Forestry, Fishery and Mining.

3. Within this category, we browse for the sectors of our interest. In this case we select "Vehicles".

4. A data source is selected from within the box, as "Conventional Pollutants".

5. In the box "Level of increased economic activity in the selected sector" introduce $11 billion.

6. Choose to have all the sectors of the economy involved though it is also possible to select the top 10, top 25, etc.

7. Hit the "Display data for selected sector" bar.

8. After a couple of seconds values depicted in Table 5.6 are shown and measured in Metric Tons.

Table 5.6 **Pollutants in metric tons (manufacturing) (backwards)**

SO_2	CO	NO_2	VOC	Lead	Particulate
3,949	6,210	3,882	1,460	3.35	678

FORWARDS EFFECT

After these cars are sold, it is necessary to also consider their operation, and the subsequent pollution they generate. To compute this pollution one has to know as an average how much gasoline these 800,000 cars will burn in one year or in other periods of time. Say for instance that statistics show that each car travels on average 15,000 km per year, and assume that on average a car yields 7.5 km/liter. From here it is possible to calculate how many million of cubic meters of fuel will be consumed by 800,000 cars during a year.

Then, using again the IO technique the amount of pollutants that these cars' operation will produce is computed, Table 5.7.

Table 5.7 **Pollutants in metric tons (operation) (forwards)**

SO_2	CO	NO_2	VOC	Lead	Particulate
3.44	5.42	3.38	1.274	0.0029	0.59

Therefore, adding up the pollution caused by the backward and forward actions, it is possible to compute the total effect the passenger car industry produces.

If we want to compute for instance its *warming effect*, this information can also be extracted from the system as follows: Values are expressed as GWP, that is the *Global Warming Potential*, which is a measure that weights the releases of these gases and converts them into releases of *Carbon Dioxide Equivalent.* Table 5.8.

Table 5.8 **Pollutants in metric tons of CO_2 E (only operation)**

GWP	CO_2 MTCO$_2$E	CH_4 MTCO$_2$E	$N_2 0$ MTCO$_2$E	CFC's MTCO$_2$E
884	799	81	0.7	2.8

Another example

For the construction of a highway there is a budget of $500 million. In this case the release of gases into the atmosphere will be as shown in Table 5.9.

Table 5.9 **Pollutants in metric tons (construction)**

SO_2	CO	NO_2	VOC	Lead	Particulate
1,054	2,145	2,274	373	0.319	2,707

Obviously, results are only as good as the information input to the model.

At a macro level, such as the selection of the options(s) to be adopted by the Federal Government to produce electric energy in the next 50 years, this technique can be used together with Mathematical Programming (see section 5.9), and then this model *will select the option which will generate the least contamination,* while complying with all the restrictions imposed. It also supplies quantitative values for criteria significance, i.e., *ranks criteria in accordance with their importance in achieving the final selection of options.*

Conclusion

The utilization of this technique solves one of the most difficult problems in EIA which is the determination of the *cumulative contamination* generated for the production of a certain good. It is believed that no other technique can reach these results. Naturally, although its operation is simple, it requires a lot of information, and it is doubtful how many countries can produce it. Nevertheless employing the data provided by Carnegie Mellon University Green Design Initiative, a fairly good calculation can be made.

Of course, the IO model, because of its complex structure, has some drawbacks, one of them being the degree of *aggregation of the economy* in a country. This is understandable when one considers that it is almost impossible to keep track of every single activity, and, as a consequence these are grouped in categories, but naturally this produces a degree of *aggregation that sometimes hides the true nature of the activities included.* Just to give an idea, consider that the IO matrix in the USA includes 500 activities, while the European Commission takes into account only 79 (European System of National and Regional Accounts) (ESA). Another restriction in the IO models is that all inter-industrial relationships are considered linear, which could not be true.

5.6 Life Cycle Analysis (LCA)

Description

This important technique analyzes as its name implies the whole life of a product, from "cradle to the grave". In so doing it determines all the inputs for manufacturing that product starting with raw materials, energy, water, etc, continues with parts made from these raw materiales, and goes on with the components made from parts (sub-assemblies and assemblies), until the final product stage is reached.

In the same way that the main branch of a river is formed with the flows of many tributaries, in making a product or executing a project there is a flow not only of raw materials but also in fuel and energy to build something. The production of these inputs *causes environmental impacts which are appraised by LCA.*

In order to list all the components, parts, sub-assemblies and assemblies, it uses a supply chain pyramid diagram which has in its *first or lowest level* all the basic input as the mentioned raw materials, water, air, fuel, etc. See Figure 5.1.

The *second level* is formed with parts or components made out of items in the first level, and this circumstance is indicated by arrows. The quantity of each component in number, kg, liters, or in any other unit of measure is indicated on the arrow.

At the *third level*, components are grouped in a similar way to form sub-assemblies, and again, on the respective arrow is shown the number of components or parts that are used in each sub-assembly. The *last level* is the finished product where all the assemblies participate in the required quantity.

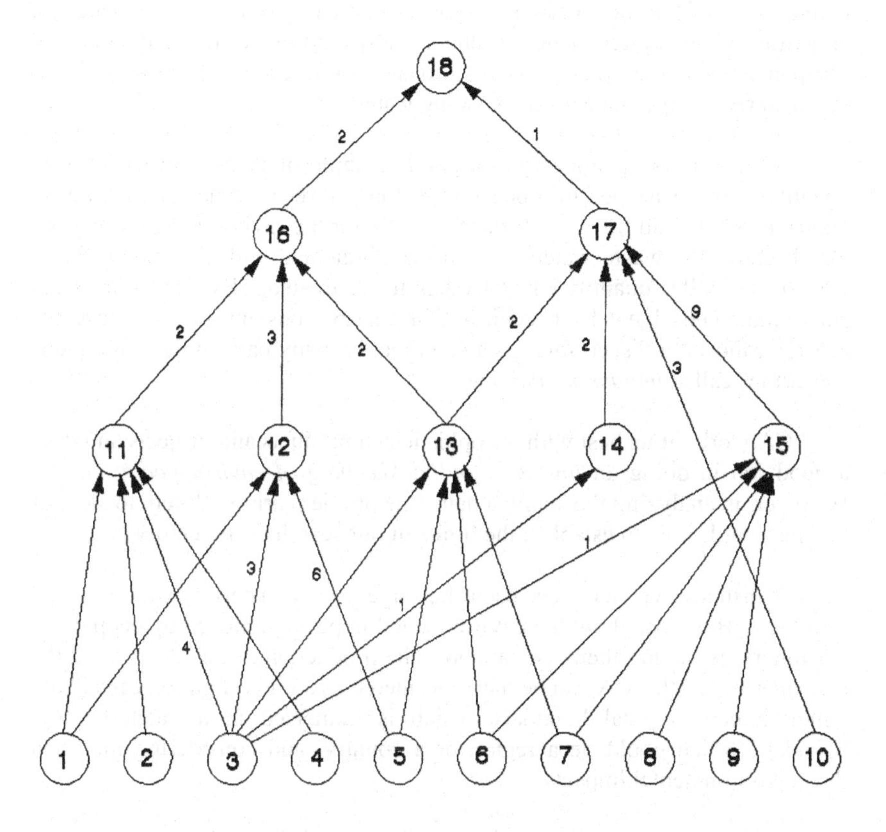

Figure 5.1. **Supply chain pyramid**

Evidently there can be as many levels as necessary, and as complicated as required.

The same procedure applies to projects, since these are composed by supporting works, sub-projects, complementary projects, etc. Any project can then be dissected and each part thoroughly and methodically analyzed. For instance a project that calls for the construction of a cement plant can be broken down in its components, e.g., rotary kiln, fuel system, crushers, packing, etc., and determine for each one its contribution to environmental impacts.

Obviously, once the plant is in operation, the volume, flow and composition of the air emissions becomes known, and as a consequence an appraisal of its effects can be done. LCA permits a rational study of components and sometimes it allows for an improvement in the quantity of raw materials usage, especially including water.

The meat packing industry is a good example of this, for the necessary quantity of water needed to process a cow has in recent times being reduced almost to a half, and this is important not only for the savings in raw water but also because the waste water is highly contaminated with fat, blood, feces, etc., so a smaller quantity of water has to be treated. By the same token substantial gains have been achieved in energy consumption. Because the process follows a discernible path or in reality many paths, this technique is sometimes called *impact path analysis.*

LCA works at its best with unsophisticated products and projects and does a good job in doing an *inventory or bookkeeping of environmental effects.* However, in analyzing the supply chain, one problem arises: Where to stop, or in other words how to establish the limits of our search or inventory?

For instance we can even reach the stage of raw material extraction, and stop there. But what about the environmental impacts produced by people and machinery extracting them? What about the production of metals to build the machineries, etc.?. LCA can be then considered as an investigation carried out within a certain local boundary which is immersed in a much broader boundary which could be a region or a country, and considering the most direct environmental impacts.

For this reason, LCA is employed together with IO analysis to determine impacts, because IO has the capability to analyze *all* direct, indirect or secondary consequences of the project, that is it can explore the *whole supply chain,* and not only the most direct links, so the boundary is now the *whole economy.* When calculations are made and compared with LCA, IO analysis result differ since it considers the economy as a whole.

Therefore there is a symbiosis between these two systems and a nice complementation, and since IO analysis investigate the monetary inputs for an industry to produce also a monetary output, this association links economic with environmental issues.

Investigators Sangwon Suh and Gjalt Huppes have proposed a new approach in the EIO-LCA symbiosis which is called HEIO-LCA, the "H" standing for *hybrid.* HEIO-LCA is based on processes rather than on prices. One of their reasons is that it is a recognized fact that the relationship between consumers and producers, that is the supply chain, is not a linear but a circular one because of recycling. Please see *"European production, consumption and waste processing modeled as integrated hybrid analysis for sustainable decision making"*, by Sangwon Suh and Gjalt Huppens at: http://www.leidenuniv.nl/interfac/cml/ssp/projects/eudgresearch/eurodecide.pdf

5.6.1 Case study - Construction of a copper concentration plant

Background information

A project, whatever its nature, involves the use and consumption of equipment, components, parts, raw materials, energy, water, minerals, vegetal fibers, etc., and of course, manpower. Take a project, say the construction of a copper concentration plant. This project will require:

- Manpower for its design, construction, financing and operation;
- Material for its construction, that is, portland cement, steel, aluminum, and already constructed inputs such as elevators, stairs, trusses, beams, columns, sidings, etc;
- Crushers to pulverize the ore;
- Equipment, for the metallurgical process, such as tanks, pumps, agitators, piping, blowers, etc;
- Chemicals;
- Transportation equipment, such as trucks;
- Synthetic fibers such as polystyrene for the lining of the tailing ponds.
- Etc.

So, its construction and operation demand - --besides manpower --- many components that when broken down to their initial input, lead to natural resources, i.e., iron ore to manufacture crushers, rails, beams, trusses, etc., alumina to make aluminum sidings for buildings, calcium carbonate and other minerals for the manufacture of portland cement, steel, glass, plastics for the manufacture of components, and so on. Therefore each manufactured component, like aluminum sheets for building sidings, starts with an initial input of alumina and lots of electric energy, water, air, etc.

The production of each of these components spews emissions to the atmosphere and to the ground or may produce discharges to bodies of water. These are the *environmental effects* we have to consider when analyzing a project. The consumption and pollution do not stop here. At the end of its economic and technical life, a plant is usually dismantled, its equipment and trusses sold as scrap metal, the glasses of their windows go to a glass smelter, the aluminum recycled, and so on; therefore, there is another contamination problem here for the *disposal and processing* of these components.

Needless to say, this inventory considers how much of a certain raw material and how many components enters in the production of one unit of one intermediate set. In turn, it is also necessary to quantify how many units of this set go in the final unit. For instance in a very simplified version and taking as an example the fabrication of a truck, it starts with the raw material (iron ore) to produce pig iron in the blast furnace and steel in the Bessemer converters. Of course the amount of iron ore, coke and other components needed to produce one ton of steel is known.

The amount of steel needed for the chassis of a heavy duty truck is obtained from the truck manufacturer's records. Technical records also specify that there is only one chassis per truck. The same applies to glass, tires, aluminum, plastic, that participate in the construction of the truck. If we consider tires for instance, it is known how many tires and what size are needed for each truck.

Now, if the calculation is made for the pollution generated to manufacture one tire, we can easily compute how much pollution corresponds to each truck. The same applies to other components. Therefore, we get a pyramid diagram where the raw materials occupy the lowest level at the bottom of the pyramid. The manufactured products go to a second level, the components to the third, the assemblies to the fourth and finally the final product, the truck, is in the fifth level. There could be many other levels and the diagram can become really complex, although the process can also be represented, more efficiently, in a stepped matrix (see section 4.3.9).

Finally, once the pollution released by the main components is known, it is a simple mathematical procedure to compute the total pollution released in the construction of a truck. If this figure is multiplied by the number of trucks in the fleet, the total amount of raw materials and components for the transportation equipment for this project can be found.

Using the dollar value for the fleet of trucks, and utilizing the software developed by Carnegie Mellon University Green Design Initiative. (see section 5.5.1), the total amount of contamination can be determined. The contamination is measured as metric tons of SO_2, CO, CO_2, NOx, and particulates.

The same procedure applies to other areas of the project (buildings, crushing equipment, production equipment, disposal equipment, etc.), and adding them up we can get the cumulative impact produced for the construction of the copper concentration plant. Now we have to do the same for the operation phase. Here, we compute the emissions released by the trucks in a working day, the pollution created in manufacturing the chemicals used by the plant, the pollution caused to generate electricity for the operation, and so on.

Perhaps the most difficult part is to determine the pollution produced in each process, and each level, say for instance in the manufacture of pig iron.

5.6.2 Case study --- Determination of contamination produced by a metallurgical mining process

Background information

To illustrate the technique, an actual project built in the United States for a mining company, is described.

The company was engaged in gold production with ore extracted from an open pit. Because the ore contained traces of some undesirable chemicals which were greatly influencing the floatation process to obtain gold, it was necessary to get rid of these chemicals. The chosen way was to pulverize the ore, after it has been roughly crushed to an appropriate size, using a rotary furnace and an in-built ball crusher. The resulting material, finely divided, was then classified through cyclones and then transported to the floatation vessels.

Gases emitted from the heating and grinding processes were treated to recover arsenates, mercury, sulphuric acid, and other chemicals. The balance of gases were then released to the atmosphere. In Figure 5.2 the process is sketched in a flow diagram at level 3, while the other levels show the existent relationships starting with raw material at level 1.

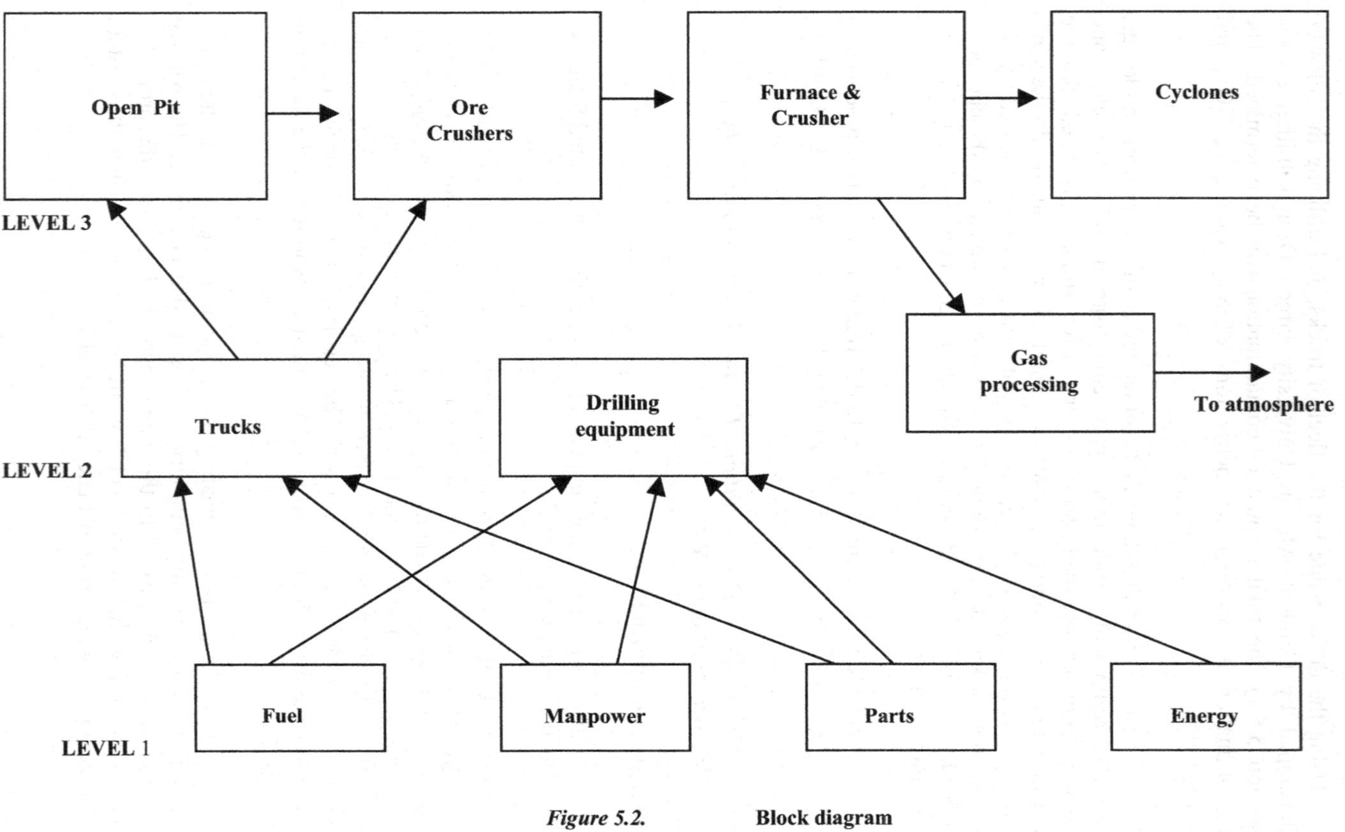

Figure 5.2. Block diagram

Therefore as per level 3:

- Ore is mined from the open pit, and then fractured using ore crushers (jaw crushers) to an adequate size.
- The resulting ore is pulverized through a rotary-ball crusher into a finer powder and at the same time heated (furnace & crusher).
- The heating process vaporizes chemicals contained in the ore, which are then piped to a gas processor. From here chemicals are extracted. Finally the remaining exhaust gas is vented to the atmosphere.
- The pulverized ore from the furnace-crusher goes to cyclones, and is then conveyed to the floatation process to get metallic gold and silver.

For the other levels:

- Primary level pertains to fuel, manpower, parts and energy.
- Secondary level corresponds to drilling equipment and the use of trucks (heavy component in this type of operation), and used to transport ore from the open pit to the jaw crushers, and from the jaw-crushers to the furnace-ball crusher.

In this process, manpower, electricity, fuel and parts are input as raw materials. Table 5.10 shows that fuel is used for trucks, for the drilling equipment and for the furnace. Manpower participates in trucks (drivers), drilling equipment, mining, and in crushers and cyclones operation. Trucks transport ore from the open pit to the jaw crushers, and from here haul the material to the furnace. Some inputs from the first level go to both the second and the third level as in the case of fuel (albeit not shown in Figure 5.2).

Others, like energy go to the drilling equipment and also directly to the jaw and ball crushers (for the electric motors), to the furnace (for the furnace rotation motors), and to the cyclones (for the electric motors). The same happens with manpower. It can then be seen that for producing one ton of pulverized ore, and not considering chemicals, different inputs are needed such as, fuel, manpower, parts and energy, intervening in different quantities to feed levels 2 and 3.

Table 5.10 **Computation and evaluation of inputs**

	Fuel	Person-nel	Parts	Energy	Trucks	Drilling equip-ment	Open pit	Ore crushers	Furnace and crusher	Cyclones	Total per day	US$ Cost
Fuel	1				1,543	76			18,400		20,019	4,004
Personnel		1			30	28	62	22	10	9	161	7,601
Parts			1		250	90		150	1,200	75	1,765	1,765
Energy				1		48,000		360,000	441,600		849,600	21,240
Trucks					1						1	
Drilling equipment						1					1	
Open pit							1				1	
Ore crushers								1			1	
Furnace and crusher									1		1	
Cyclones										1	1	
											TOTAL DAILY	34,610
											TOTAL ANNUAL	12,632,650

We need to know how much fuel is required to produce the whole plant output, how many people are needed, the dollar value in parts and the total energy comsumption, etc. Table 5.10 shows these quantities. For instance, in the fuel row, 51.4 liters are needed per truck and per day. If 30 trucks are required, then the consumption will be 51.4 x 30 = 1,543 liters of fuel per day, just for the trucks.

But fuel is as well used for drilling equipment in an amount of 76 liters/day, so, this amount has to be added to the figure just computed for trucks. Since natural gas is also employed for the furnace, with a fuel equivalent of 18,400 liters per day, the total quantity of fuel is 20,019 liters per day. This magnitude, multiplied by the average price of fuel, gives the whole daily amount of US$ 4,004. The same procedure is followed for the other inputs, and a final cost of US$ 34,610 per day is found.

Finally, multiplying this cost for the number of working days in a year we get a figure of US$ 12,632,650. With this value we use the Carnegie Mellon University Green Design Initiative method (see section 5.5.1) and obtain the figures shown in Table 5.11 for contaminants released to the atmosphere for this production. The procedure to get these values is as follows:

1. Enter the Carnegie Mellon University Green Design Initiative (2003) at their www.eiolca.net web site.
2. Browse under "Categories" for the area of our interest. In this case it is Forestry, Fishery and Mining.
3. Within this category, browse for the sectors of interest. In this case select :"Copper ore".
4. A data source is selected from within the box, such as "Conventional Pollutants".
5. In the box of "Level of increased economic activity in the selected sector" we introduce in millions of dollars the above determined value, that is: 12.632.
6. We choose to have all sectors of the economy involved but it is also possible to select the top 10, top 25, etc.
7. We hit the "Display data for selected sector" bar.
8. After a couple of seconds values depicted in Table 5.11 are shown and measured in metric tons.

Table 5.11 **Contaminants release**

SO_2	CO	NO_2	VOC	Lead	Particulate
MT	MT	MT	MT	MT	MT
135	62.7	95	13.7	0.06	72.6

Therefore these are the pollutants and their amount generated by the operation of this metallurgical process.

Conclusion

LCA shows the existent relationships between a project and the pollution it creates through its whole life. If the project has an economic objective, then we can have a relationship between economic benefits and sustainability, or put in another way, it is *possible to evaluate how much in terms of environmental costs we are willing to pay for a certain economic development.*

5.7 Multicriteria Analysis (MCA)

Description

In some way or the other all the techniques try to *assess a project or projects against a set of criteria.* The denomination of multicriteria as well as multiattributes comes from this circumstance, and these tools are used for the analysis of projects, plans, programmes and options either with a single objective or with several, and as mentioned with many different attributes or criteria.

These techniques attempt to solve problems with *different objectives which normally are opposed,* such as the classical example of minimizing the environmental cost and at the same time maximizing the economic development, and they are the methods that are receiving the preferred attention in numerous countries especially The Netherlands and the U.K.

In the Americas, Canada and the USA there is an increasing trend in using this type of technique. One of the advantages of this methodology is that it not only can work with weights for projects and for criteria, but too with one or with many projects or alternatives. Some of them, like the Analytical Hierarchy Process does not yield a unique solution but a *prioritized set of projects or alternatives*. This type of result constitutes a very useful guide for stakeholders and decision makers since it *provides the elements conducive to an educated decision.* This technique is used all around the world for many kinds of problems, where EIA is one of them.

Other techniques such as Mathematical Programming (MP), operate in a different way and are able to provide an *optimal solution,* and as a consequence the *best selection out of a listing of competing projects* or *alternatives,* as well as, if so desired, a ranking of these competing projects. When funds are scarce it is very important to allocate those resources in such a way as to optimize their use, and for this, *MP is the tool of choice.* MP can work with or without weights, and with any kind of criteria and units of measure. Besides, it provides very important information regarding criteria significance, through the calculation of the marginal cost.

A very important concept that is ingrained in MP is the *computation of trade-offs between projects, among criteria and between projects and criteria*, because the essence of this matrix method relies precisely in the computation of trade-offs, and in selecting the best set of them which optimizes the proposed objective. We are not saying that one of the two discussed methods is better than the other. It is believed that AHP has definite advantages regarding comprehension and ease of use, and with a very understandable commercial software.

MP on the other hand is more difficult to understand especially in its mathematical foundation, and it does not have a dedicated software for its resolution, but utilizes add-ins that come in advanced spreadsheet programs, and also can utilize hundreds of either commercial or free software, which are called codes.

In this book all examples using MP have been solved using the "Solver" add-in included in Excel®. MP is utilized everyday around the world for many applications other than for EIA, and it is said that 80 % of Fortune 500 companies employ MP in one way or the other. As a conclusion it is believed that for EIA applications, MP offers some advantages over AHP, especially when considering the application of *thresholds for criteria.*

Objectives

Most common objectives employed by these techniques are:

- To achieve a minimum cost or to maximize a benefit;
- To maximize employment;
- To maximize competitiveness;
- To optimize distributional impacts (that is, impact in certain sectors).

In the case of multiple objectives, a technique that belongs to the MP family and called "Goal Programming" can handle this type of situation. Goal Programming is not included as an add-in in Excel® or in other spreadsheet programs, but a stand-alone version of "Solver", in a commercial format includes Goal Programming.

The Multicriteria Analysis (MCA) has proved to have definite advantages over other methods for appraisal, since it can manage problems that are not strictly related to economic issues, so they are more appropriate for EIA purposes, as many of them can incorporate qualitative as well as quantitative information. From now on, methods with increasing complexity will be studied. They are:

5.7.1 Dollar value appraisal

Here, the objective is to make a selection of projects or alternatives, subject to different criteria, but based on *how much value is obtained* with *one dollar invested in a project.*

5.7.1.1 Case study --- Selection of different urban projects

Description

Consider for instance the following example: A City Hall has a portfolio of nine different projects , related to different fields, like social, environment, infrastructure, population density, etc. The costs associated with each project are known, as detailed in Table 5.12.

Table 5.12 **Cost Table**

Project number	Project	Cost [000's US$]
1	Water Treatment Plant upgrade	2,840
2	Repair hwy. 23	8,912
3	Repair water main	4,380
4	Construction of storm water drains	3,650
5	Improve density	5,263
6	Increase water supply	4,224
7	Reforestation	4,664
8	Construction of new houses for low income people	7,210
9	Improve sewage	6,250
	TOTAL:	**47,393**

A selection of projects has to be made considering a set of criteria as follows:

A. Cost
B. Relationship to strategy

C. Return on investment
D. Multiplier effect
E. Public safety
F. Public health
G. Environmental impact
H. Community support

Each project has been appraised with each criterion, through a panel of experts that produced the values (scoring) indicated in Table 5.13. It can be seen for instance that the intersection of the column for project "Repair water main" with the row corresponding to criterion "Environmental Impact", has been graded as 0.65, on an arbitrary scale. The higher the value the more important. There are also values for the same criterion related to all other projects. Notice that a certain project can have the highest value for one criterion and the lowest value for another.

Also from expert advice a weight has been assigned to each criterion which is depicted in the left column of Table 5.13. In order to reflect the importance of the criteria weights, each value from Table 5.13 is multiplied by the corresponding weight. Thus, for instance for the project "Increase water supply", the value at its intersection with row "Multiplier effect" (0.70), is multiplied by the "multiplier effect" weight (criterion D), which is 14. Therefore this project gets a score of: 0.70 x 14 = 9.80. The same procedure applies to all values and Table 5.14 is obtained.

Remember that a column corresponds to a project, and that the values in that column *represent the appraisal of the importance of this project related to each criterion*. The summation of the values in the column of a certain project, say "Repair Road" is 70 (Table 5.14), and will indicate its importance when compared with others. If now each one of these totals is divided by the cost of the corresponding project, the quotient will indicate the dollar value of such a project. See "Dollar value" row in Table 5.14, and notice that a ranking of projects can be obtained from these values.

It can be seen in Table 5.14 that the project that gives the most value for dollar is project number 4 "Construction of storm water drains", followed by projects number 6, 9 and so on. This is a very simple procedure to rank projects according to their contribution to the objective.

Table 5.13 **Projects scoring as per criteria**

		Project numbers	1	2	3	4	5	6	7	8	9
			WTP upgrade	Repair road	Repair water-main	Construc-tion of storm-water drains	Improve density	Increase water supply	Reforest-ation	Construc-tion of houses for low income people	Im-prove sewage
		Project cost (000's)	2,840	8,912	4,380	3,650	5,263	4,244	4,664	7,210	6,250
Criteria	**Description**	**Weights**	Performance matrix								
A	Cost	*25*	0.20	0.20	0.80	0.90	0.40	1.00	0.15	0.65	1.00
B	Relationship to strategy	*20*	0.45	0.85	0.25	0.20	0.90	0.45	0.85	0.10	0.25
C	Return on investment	*25*	0.35	0.30	0.30	0.80	0.60	0.60	0.70	0.10	1.00
D	Multiplier effect	*14*	0.90	0.10	0.05	0.50	0.20	**0.70**	0.60	0.60	0.30
E	Public safety	*18*	0.05	0.20	0.90	0.10	0.90	0.70	0.60	0.40	0.55
F	Public health	*22*	0.05	0.95	0.15	0.25	0.65	0.10	0.20	0.35	0.80
G	Environmental impact	*20*	0.45	0.15	**0.65**	0.10	0.80	0.25	0.80	0.30	0.65
H	Community support	*15*	0.25	0.80	0.80	0.55	0.10	0.25	0.35	0.40	0.80

Table 5.14 **Projects scoring as per criteria (weighted)**

		Project numbers	1	2	3	4	5	6	7	8	9
			WTP upgrade	**Repair road**	**Repair water-main**	**Construc-tion of storm-water drains**	**Improve density**	**Increase water supply**	**Reforest-ation**	**Construc-tion of houses for low income people**	**Im-prove sewage**
		Project cost (000's)	**2,840**	**8,912**	**4,380**	**3,650**	**5,263**	**4,244**	**4,664**	**7,210**	**6,250**
Criteria	**Description**	**Weights**	**Performance matrix**								
A	**Cost**	***25***	5.00	5.00	20.00	22.50	10.00	25.00	3.75	16.25	25.00
B	**Relationship to strategy**	***20***	9.00	17.00	5.00	4.00	18.00	9.00	17.00	2.00	5.00
C	**Return on investment**	***25***	8.75	7.50	7.50	20.00	15.00	15.00	17.50	2.50	25.00
D	**Multiplier effect**	***14***	12.60	1.40	0.70	7.00	2.80	**9.80**	8.40	8.40	4.20
E	**Public safety**	***18***	0.90	3.60	16.20	1.80	16.20	12.60	10.80	7.20	
F	**Public health**	***22***	1.10	20.90	3.30	5.50	14.30	2.20	4.40	7.70	9.90
G	**Environmental impact**	***20***	9.00	3.00	13.00	2.00	16.00	5.00	16.00	6.00	17.60
H	**Community support**	***15***	3.75	12.00	12.00	8.25	1.50	3.75	5.25	6.00	13.00

			1	2	3	4	5	6	7	8	9
	Sum		50	70	78	71	94	82	83	56	112
	Dollar value		1.76	0.79	1.77	**1.95**	1.78	1.93	1.78	0.78	1.79
	Ranking		7th	8th	6th	1st	5th	2nd	4th	9th	3rd

Table 5.15 **Solution by Mathematical Programming**

Criteria	Project numbers	1	2	3	4	5	6	7	8	9			
		WTP upgrade	Repair road	Repair watermain	Construction of stormwater drains	Improve density	Increase water supply	Reforestation	Construction of houses for low income people	Improve sewage	From computation	Sign	Thresholds
	Project costs (000's)	2,840	8,912	4,380	3,650	5,263	4,244	4,664	7,210	6,250			
Criteria	Description	Performance matrix											
A	Cost	5.00	5.00	20.00	22.50	10.00	25.00	3.75	16.25	25.00	38	<	133
B	Relationship to strategy	9.00	17.00	5.00	4.00	18.00	9.00	17.00	2.00	5.00	27	<	86
C	Return on investment	8.75	7.50	7.50	20.00	15.00	15.00	17.50	2.50	25.00	34	<	119
D	Multiplier effect	12.60	1.40	0.70	7.00	2.80	9.80	8.40	8.40	4.20	15	<	55
E	Public safety	0.90	3.60	16.20	1.80	16.20	12.60	10.80	7.20		24	<	79
F	Public health	1.10	20.90	3.30	5.50	14.30	2.20	4.40	7.70	9.90	28	<	77
G	Environmental impact	9.00	3.00	13.00	2.00	16.00	5.00	16.00	6.00	17.60	24	<	83
H	Community support	3.75	12.00	12.00	8.25	1.50	3.75	5.25	6.00	13.00	21	<	65
	Sum	50	70	78	71	94	82	83	56	112			
Funds available (000's)		1	1	1	1	1	1	1	1	1	15,313	<	15,798

RESULT (Selected projects identified by "1s")

1	0	0	0	1	0	0	1	0

5.7.2 Dollar value with monetary restrictions

Let us consider again the example proposed in the last section. As stated the total cost of all these undertakings is $ 47,393,000. Assume, as is most frequently the case, that City Hall does not have such an amount of money, but only about a third of it, i.e., $ 15,798,000. The problem is now to select projects as before, but with the added restriction of not having enough funds.

In this case, and ranking the projects as per their dollar value, the only projects selected will be those that can be executed with that limited budget. It is simple enough, following the ranking to realize that projects 6, 4, and 9, foot the bill with a total amount of $ 14,144,000, and leave an unused remnant of $ 1,654,000. In this case the available funds are used is 89%. There is no doubt that the simple scheme depicted above can't consider these type of situations and then we have to look for more elaborate methods.

Is there any way to improve this selection and get more value for our money? The answer is yes.

If MP is used, it is found (see Table 5.15) that instead of projects 6,4,9, now the model chose projects 1,5,8, which make a better use of the scarce available funds. Now the total cost for the three selected projects is $ 15,313,000 with a remnant of $ 485,000 so 97% of the available funds are used.

5.7.3 Environmental damage appraisal

In example 5.7.1.1 all the values in Table 5.13 were obtained from expert opinion or from *public consultation.* However, in most of the cases there are *deterministic values* such as *content of pollutants* in $\mu g/m^3$, or *flows* measured in m^3/sec.

In these cases the elementary procedure does not work because quantities with different units cannot be added. But this is indeed not a problem for it is always possible to make a *normalization.* There are several ways to perform a normalization, but one of the simplest consists in expressing each value, for each row, as a percentage of the largest value in that row.

This way percentages are obtained that can be added, and the appraisal can be expressed as an environmental damage value instead of a dollar value.

5.7.3.1 Case study - Selection of a chemical process to minimize environmental costs

Background information

An industrial firm has two chemical industrial processes to manufacture the same product, with different costs for each process, and delivering the same output in tons of finished products. It is assumed that the plant will be located in the outskirts of a city. The different alternatives are shown in columns in Table 5.16 and identified as A and B. In order to select alternative A or B the appraisal is made with respect to their *respective contribution to contamination of the environment,* and the objective is to select the alternative *producing the least pollution*, and in this case, regardless of its cost.

As a *measuring device,* nine different criteria are utilized, concerning pollutants produced by the two different processes. Also, through expert opinion, a *relative weight* has been assigned to each criterion, measuring the importance that each one has relative to the others. Each process creates different kinds of environmental impacts as follows:

Water impact:

- Discharge of *hot water* into a lake, which modifies the fish habitat and produces algae;
- Discharge of *treated water* with a high content of BOD_5, reducing the oxygen content in the water.

Atmospheric impact:

- NOx, CO and SO_2 are released to the atmosphere. They have an effect in climate change and the last one can produce respiratory problems and damages to structures in the city;
- Particulates are also spewed by the two processes.

Land impact:

- Whatever the choice of processes they will be using agricultural land, surrounding the city.

Noise impact:

- Both alternatives will generate the same level of noise.

Energy:

- Consumption of energy is detailed for each plant.

The objective is to select the process that will cause the *least damaging effect to the environment.* Since there is a range of units of measure it is not possible just to add the values up, so the first step is to normalize or *homogenize* them by applying a simple rule: For each criterion select the maximum figure, assign it a 100 value, and calculate the other value on this criterion respective to the highest.

Thus, for criterion *Discharge of heated water to the lake* for instance, there are two values: 457 m^3/day for alternative A and 721 m^3/day for alternative B. So, choosing 721 m^3/day as value 100, the other one is (457/721) x 100 = 63.

This normalized data is reproduced in Table 5.17.

Table 5.16 **Data on chemical processes**

ORIGINAL DATA

Matrix (A)

	Criteria weights	A	B	Units
Process		**A**	**B**	
Cost (million of US$)		**78.623**	**72.928**	
Discharge of heated water to the lake	*8*	457	721	m^3 / day
BOD_5 production	*10*	2,400	1,900	ppm
NOx production	*18*	0.080	0.070	mg / m^3
CO production	*16*	2.950	2.420	ppm
SO_2 production	*7*	0.050	0.060	mg / m^3
Particulate production (< 10 microns)	*18*	0.150	0.154	mg / Nm^3- day
Land use	*10*	42	43	hectares
Noise generation	*2*	10	10	decibels
Electricity consumption	*11*	1,562	1,921	kWh /year

Table 5.17 **Normalized data**

Matrix (B) (Normalized)

Values from Matrix (A)

Computed as percentages over the highest values in each row

Process	A	B
Discharge of heated water to the lake	63	100
BOD_5 production	100	79
NOx production	100	88
CO production	100	82
SO_2 production	83	100
Particulate production (<10 microns)	97	100
Land use	98	100
Noise generation	100	100
Electricity consumption	81	100

Values for Table 5.18 are obtained by applying the criteria weights to the values of Table 5.17.

Table 5.18 **Normalized and weighted data**

Process	A	B
Discharge of heated water to the lake	507	800
BOD_5 production	1,000	792
NOx production	1,800	1,575
CO production	1,600	1,313
SO_2 production	583	700
Particulate production (<10 microns)	1,753	1,800
Land use	977	1,000
Noise generation	200	200
Electricity consumption	894	1.100
Damage to the environment	9,315	**9,279**
Damage per dollar	**0.118**	0.127

Summation of values for each alternative indicates that alternative A accumulates a damage to the environment "valued" at 9,315 (a non-dimensional figure), and for alternative B, a figure of 9,279. As a consequence it shows that alternative B is better than alternative A from the *environmental point of view.* A cost factor could perhaps be introduced in the analysis, dividing this damage in "marks" or "points" by the construction costs (damage per dollar). In this way it is possible to have an idea of how much damage is caused by one dollar in each process.

This example shows that one dollar in alternative A causes a damage of 0.118 "damaging units", which is lower than the value produced by alternative B, so from this point of view it appears that A is better than B. The problem here is that both objectives cannot be simultaneously considered, that is *minimizing damage to the environment* and *minimizing damage per dollar*, so a decision has to be made, and be justified.

Introducing more factors

In our analysis two approaches have been taken into account. The first one, focuses on selecting the alternative that produces de least environmental damage. The second one, relying on dollar value, selects an alternative according to the principle of getting the best value for money. Both approaches are based on predetermined criteria to gauge the contribution of each alternative to achieving the objective. But in reality, each approach, albeit useful in some situations, is too simplistic in others.

Thresholds

In the last case, assuming for instance that NOx production is lower in one alternative than in the other, does not necessarily mean that it is permissible, since in each case NOx production could be exceeding some safety limits. The same happens to all the other criteria, except perhaps for land use and energy. Besides, some criteria are expressed as units of production, but it could be the case that there are criteria where it is not a matter of production but of capacity, or some other issue.

As an example, assume in the last example that each process will produce treated non-contaminated waste water in very different quantities that can be safely discharged into the city sewage system. Here there are no reservations about the quality of the water discharged, or its temperature, but there is another type of concern. Is there a certainty that the city water treatment plant *will have the capacity* to handle this additional load?

So, we need a "capacity criterion", and if a maximum value as a capacity limit is assigned to this criterion, then the solution will have to consider this circumstance in selecting one of the processes, or perhaps rejecting both.

In other cases, even if there are quantitative values for scoring, it does not mean that these values are exactly known. What if they are *probabilistic*, such as in the case of some sort of risk?.

Or suppose for instance that one of the projects in the strategic plan for a city calls for a housing construction program for low income people. We do our homework through a survey and determine that 25 % percent of the population in a certain area is composed of up to three members in a family, that 62 % is composed of four members in a family, and that 13 % is composed of more than four members in a family.

Remember, these are not exact values and there is a range for each of the three categories, so how do we manage in this case? Something similar happens to certain types of *risk in projects*. The answer lies in the fact that it is necessary to work, for each criterion, with *some kind of limiting values,* or with a *range* of them. These limiting values are called *thresholds,* and their importance is paramount in sustainable development (see section 1.9), and in the concept of carrying capacity (see section 1.11).

Cumulative effects

What other issues are to be considered in the analysis? Suppose that a country is working on implementing a policy to reduce contamination from cars. Of course, this a very realistic example since many countries or even states, such as California, are pursuing that aim.

There are several alternatives that have been explored such as:

1. More efficient gasoline engines to produce contamination below a certain threshold;
2. Cars equipped with hybrid engines, i.e., a combination of a gasoline engine and an electric motor;
3. Fuel cells;
4. Electric cars;
5. Etc.

Each one of these alternatives and others are carefully studied and weighted. However, this research process most often is very limited in *geographical space and time.* What does this mean?

It means that people are analyzing different alternatives applied to a physical space (a country, a province, a county, etc.), and to a temporal scale (now), even if the consequences of the policy are likely to last for decades to come. What is incomplete in this approach?

A primary omission is that none of the plans allow for input from past activities that will affect not only present day decisions but the future as well. For instance, most people agree on the convenience of "zero emission" vehicles, using either electric batteries or fuel cells, but few consider that to manufacture a battery one needs inputs such as cadmium, lithium or some other metal, as well as various chemicals. Of course, there will be zero emission during the operation of this electric car, but what about the fuel needed to produce the energy to charge the batteries?

The same happens with fuel cells. The output of a vehicle equipped with them will be harmless emissions and water. But wait a minute, harmless emissions? There are different types of fuel cells, probably the most promising is the system using methanol as a fuel, the DMFC type (Direct Methanol Fuel Cell). Most of the fuel cells produce clean water, but also CO_2, a *greenhouse gas*, to say nothing of the emissions by the process needed to manufacture methanol, or to extract hydrogen from some source. Granted, even with the production of some CO_2 , the environment will be much cleaner than with the harmful system existent today with gasoline engines.

The concept is that, in choosing a fuel policy that will be an alternative to the current contamination by automobiles, *one must also consider what it takes to manufacture their elements (the past), as well as the inputs for their operation (the future).* As expected, this type of analysis cannot be carried out with the two simple schemes that were discussed above. For that something much more powerful is needed, and one of the techniques to deal with this subject is called "Life Cycle Analysis" (see section 5.6), which is another factor that has to be *introduced as a criterion in the multicriteria analysis.*

5.8 Analytical Hierarchy Process (AHP)

Description

This method was created by the American mathematician Thomas Saaty. The main aspects of this technique are:

a. The use of pair-wise comparison matrices, to compare criteria between themselves as well as projects or alternatives between themselves, and using a system of preferences;

b. With the values of these comparisons a mathematical procedure is applied, finding the *eigenvalues and eigenvectors* for these matrices (see Appendix, section A.1.6). Eigenvectors are then used to determine the *weight of each criteria;*

c. Values obtained from the pair-wise comparison *between alternatives* are then affected by the criteria weights;

d. The final result shows the ranking of alternatives represented in a column vector called Global Priority.

5.8.1 Case study --- Selection of renewable energy sources

Background information

A community has decided to use the renewable sources of energy available in their territory, to complement the electricty they receive from an electrical utility. A pre-feasibility study finds three potential types or options:

- Type 1: Mini-hydro units;
- Type 2: Biomass;
- Type 3: Wind energy.

There are convenient financial conditions for equipment as well as a preferential interest rate of government loans under the Clean Air Act. Figure 5.3 shows the *hierarchy structure* for this case, but, for the sake of clarity, with only one relationship between devices or type of equipment and a criterion, although the calculation involves all criteria.

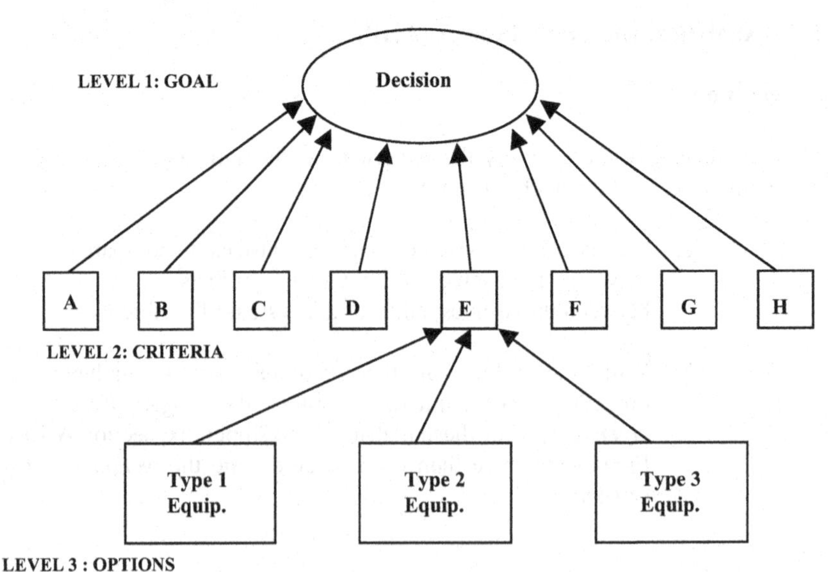

Figure 5.3. **Hierarchy structure**

Criteria used

Durability (A):
Requirement that the equipment be reliable for a long period of time.

Noise level (B):
This is data published by each manufacturer in decibels.

Ability to pay (C):
People ability to pay yearly for each type of equipment, considering a 4 % rate of interest and a total period of three years for repayment. If the community is using their own funds, the interest rate should represent what that money would earn if put to other use (cost of opportunity).

Environmental impact (D):
Estimate of the impact of each piece of equipment.

Land use (E):
Expresses the perception about land use for each type of equipment and then considered as a whole. It includes service roads to the facilities.

Property depreciation (F):
The potential loss of values of dwellings due to the proximity of the installations.

Aesthetics (G):
The pleasent or unpleasant consequences from the aesthetical point of view.

Opinions from other users (H):
The community consulted with other communities about the advantages and disadvantages of each type of equipment.

Problem representation and hierarchy

A decision has to be taken on what type of equipment to choose. The selection is appraised through the use of eight criteria, therefore every type of equipment is *evaluated according to each one* of the eight criteria.

First step: Calculation of criteria weights

To compute criteria weights two of them are paired and compared using a

1 to 9 scale, in accordance with Table 5.19.

Table 5.19 **Preferences scale**

Judgment of preference	Values
Both are equally important or preferred	**1**
One of the criteria is moderately important or preferred over the other (weak preference)	**3**
There is a strong preference of one over the other	**5**
One is strongly important over the other	**7**
One is absolutely preferred over the other, or is definitely more important	**9**
Intermediate values	**2-4-6-8**

Expert opinion applies the values above mentioned and thus Table 5.20 can be built.

Table 5.20 **Criteria assessments**

	A	B	C	D	E	F	G	H
A	1	0.5	1	1	8	2	0.25	1
B	2	1	**5**	2	1	6	0.143	0.2
C	1	**0.2**	1	0.333	6	8	5	0.125
D	1	0.5	3	1	1	1	1	2
E	0.125	1	0.167	1	1	0.125	8	4
F	0.5	0.167	0.125	1	8	1	2	8
G	4	7	0.2	1	0.125	0.5	1	6
H	1	5	8	0.5	0.25	0.125	0.167	1

For instance we compare *criterion B row* against *criterion C column*, and if there is a strong preference of B over C , then this relationship obtains a mark of **5**. Conversely, the importance of *criterion C row* compared to *B column* is the *inverse*, i.e., 1/5 = **0.2**. The same procedure is applied to all pairs of criteria, and then the set of values thus obtained is called an *asymmetric matrix*. The mathematics of the AHP method is based in this circumstance (see Appendix, section A.2).

Now we need to calculate the weights for each criterion. For this it is necessary to utilize linear algebra and compute the eigenvalues and eigenvectors (see Appendix, section A.2). However, a simpler procedure can be utilized. It implies computing for each row the following formula:

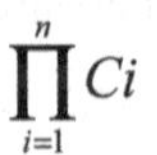

$$\prod_{i=1}^{n} Ci$$

where:
C = Criteria
n: Number of criteria

That is, for each row *all values of that row are multiplied between them.* Then for criterion A for instance, this product will be:

$$\prod_{i=1}^{n} Ci = 1 \text{ x } 0.5 \text{ x } 1 \text{ x } 1 \text{ x } 8 \text{ x } 2 \text{ x } 0.25 \text{ x } 1 = 2$$

Then the *eighth root of this value* is found, i.e. $\sqrt[8]{2} = 1.09$

Why the eighth root?. Because there are eight criteria.

When these two calculations are completed for each criterion, which is easily and accurately done in a spreadsheet, the corresponding values can be depicted in the third column of Table 5.21.

Table 5.21 **Criteria weights**

Criteria	$\prod_{i=1}^{8} Ci$	$\sqrt[8]{col._2}$	**Criteria weights**
column 1	**column 2**	**column 3**	**column 4**
A	2.000	1.09	**0.13**
B	3.432	1.17	**0.14**
C	2.000	1.09	**0.13**
D	3.000	1.15	**0.14**
E	0.083	0.73	**0.09**
F	1.333	1.04	**0.13**
G	2.098	1.10	**0.14**
H	0.104	0.75	**0.09**
	Total	8.12	

Finally, criteria weights are obtained in column 4 dividing each value in the third column by its total value of 8.12.

Second step: Calculation of alternatives or options weights

To compute weights corresponding to *types or options* every one of them is compared with the others. *But on what ground is this comparison made? By using a criterion at a time* For instance let us compare the three types of options using criterion (A) *Durability.* In this case Table 5.22 can be constructed.

Table 5.22 **Ranking alternatives with respect to criterion: *Durability (A)***

Alternatives	Type 1	Type 2	Type 3	Product	3rd root	Weight
Type 1	1	0.8	1.120	0.896	0.964	**0.303**
Type 2	1.250	1	3	3.750	1.554	**0.488**
Type 3	0.893	0.333	1	0.298	0.668	**0.210**
				Total:	3.185	

Therefore, if equipment or Type 2 is compared with equipment or Type 3 and it is found that under criterion *Durability,* Type 2 is preferred over Type 3, then in accordance with Table 5.19 a value of "3" is assigned. Since comparing Type 3 with Type 2 is its inverse, the value will be 1/3 = 0.333.

Now, alternatives or Types are again compared but using the following criterion (B), *Noise level,* and thus Table 5.23 is obtained.

Table 5.23 **Ranking alternatives with respect to criterion: *Noise level (B)***

Alternatives	Type 1	Type 2	Type 3	Product	3rd root	Weight
Type 1	1	3	0.250	0.750	0.909	**0.279**
Type 2	0.333	1	0.80	0.267	0.644	**0.197**
Type 3	4	1.25	1	5	1.710	**0.524**
				Total:	3.262	

By the same token if for instance Type 3 is compared with Type 1, a value "4" is assigned, which shows that under criterion *Noise level,* Type 3 is a little better than Type 1. The inverse, that is comparing Type 1 with Type 3 is then 0.250. The same procedure is applied to all other comparisons using the remaining criteria, and so Tables 5.24 to 5.29 are constructed.

Table 5.24 **Ranking alternatives with respect to criterion:** ***Ability to pay (C)***

Alternatives	Type 1	Type 2	Type 3	Product	3rd root	Weight
Type 1	1	1.176	0.444	0.523	0.806	**0.262**
Type 2	0.850	1	3	2.55	1.366	**0.444**
Type 3	2.250	0.333	1	0.750	0.909	**0.295**
				Total:	3.080	

Table 5.25 **Ranking alternatives with respect to criterion:** ***Environmental impact (D)***

Alternatives	Type 1	Type 2	Type 3	Product	3rd root	Weight
Type 1	1	0.852	0.430	0.366	0.716	**0.213**
Type 2	1.174	1	0.330	0.387	0.729	**0.217**
Type 3	2.326	3.030	1	7.047	1.917	**0.570**
				Total:	3.362	

Table 5.26 **Ranking alternatives with respect to criterion:** ***Land use (E)***

Alternatives	Type 1	Type 2	Type 3	Product	3rd root	Weight
Type 1	1	2	3	6	1.817	**0.532**
Type2	0.5	1	0.25	0.125	0.5	**0.146**
Type 3	0.333	4	1	1.333	1.101	**0.322**
				Total:	3.418	

Table 5.27 **Ranking alternatives with respect to criterion:** ***Property depreciation (F)***

Alternatives	Type 1	Type 2	Type 3	Product	3rd root	Weight
Type 1	1	0.555	0.333	0.167	0.550	**0.168**
Type 2	2	1	0.800	1.600	1.170	**0.357**
Type 3	3	1.250	1	3.750	1.554	**0.475**
				Total:	3.274	

Table 5.28 **Ranking alternatives with respect to criterion: *Aesthetics (G)***

Alternatives	Type 1	Type 2	Type 3	Product	3rd root	Weight
Type 1	1	3	0.890	2.670	1.387	**0.447**
Type 2	0.333	1	2.800	0.933	0.977	**0.315**
Type 3	1.124	0.357	1	0.401	0.738	**0.238**
				Total:	3.182	

Table 5.29 **Ranking alternatives with respect to criterion: *Opinion from other users (H)***

Alternatives	Type 1	Type 2	Type 3	Product	3rd root	Weight
Type 1	1	0.300	1.250	0.375	0.721	**0.233**
Type 2	3.333	1	0.345	1.149	1.048	**0.339**
Type 3	0.800	2.900	1	2.320	1.324	**0.428**
				Total:	3.092	

Third step: Application of criteria weights to alternatives

Now all the information obtained from the above Tables is consolidated and a *Performance Matrix* can be built. Table 5.30.

Table 5.30 **Performance matrix**

		Alternatives	Type 1	Type 2	Type 3
Criteria					
A			0.303	0.488	0.210
B			0.279	0.197	0.524
C			0.262	0.444	0.295
D			0.213	0.217	0.570
E			0.532	0.146	0.322
F			0.168	0.357	0.475
G			0.447	0.315	0.238
H			0.233	0.339	0.428

This performance matrix shows in each cell the *geometric mean* of an alternative when it has been compared with all the other alternatives and using a specific criterion. These values come from the *weight* column in each table.

For example: Value **0.197** in cell (criterion B and Type 2), is the geometric mean of all the values obtained when comparing Type 2 with Type 1 and Type 3, and using for that comparison criterion B.

Finally, *criteria weights* (from Table 5.21, column 4), are applied to these values, see Table 5.31. For instance for Type 2 and criterion B:

$$0.197 \times 0.15 = 0.029$$

As it can be seen, the Type 3 device has been selected as the best alternative, because it offers the *maximum value, or maximum weight.*

Table 5.31 **Alternatives ranking**

		Alternatives	**Type 1**	**Type 2**	**Type 3**
Criteria	**Criteria weights**				
A	**0.13**		0.041	0.066	0.028
B	**0.14**		0.040	**0.029**	0.076
C	**0.13**		0.036	0.060	0.040
D	**0.14**		0.030	0.031	0.082
E	**0.09**		0.049	0.013	0.029
F	**0.13**		0.022	0.046	0.061
G	**0.14**		0.056	0.056	0.030
H	**0.09**		0.022	0.032	0.040
		Alternatives ranking:	0.296	0.344	**0.387**

Conclusion

AHP offers a very comprehensive and simple mode of evaluation when *preferences* are involved. It is clearly understood by most people and gives accurate results. Obviously all these prior calculations made "by hand" had only the purpose of showing the model step by step, but there is a powerful software, called *Expert Choice* that allows one to perform all these computations fast and easily. This software does not really compute the weights for criteria and for alternatives using the geometric mean procedure as used here. It employs *eigenanalysis,* principles of which are briefly explained in the Appendix, section A.1.6.

This elegant method is very attractive because it is clear and logical, being each cell of the performance matrix a *geometric mean* of each alternative, which is then affected by the corresponding weight. However, its implementation, at least for EIA suggests some problems. They are:

a) As explained above, values for comparison come from *expert opinion and judgment*, which is an excellent procedure where no technical values are available, and *especially when only valued preferences are needed.*

Nevertheless it poses some problems.
As an example assume that seven different projects in a city compete for selection with respect to a certain goal. For the sake of argument suppose that these projects involve infrastructure, social, health, economics, and safety issues, and that they are:

1. Construction of storm water drains to stop frequent flooding in the north sector;

2. Construction of a road linking several areas of the city that are at present time not connected;

3. Construction of three health centers in three different areas;

4. Increase safety by increasing police surveillance in the east sector;

5. Implement a plan for creating new economic opportunities;

6. Improve transportation;

7. Pave 550 blocks of urban streets.

If we use a *pair-wise comparison between projects,* utilizing for instance as a social criterion *the needs of the people*, it means that the analyst has to have a thorough knowledge of *each project* related with *people needs*, in the *social, infrastructure and economic* fields, otherwise it is impossible to state that one project is preferred over the other on this criterion, and of course, the same applies for the other criteria.

It can be argued that usually several experts, each one in a specific field, are performing this task, but if one of them is in the social arena for instance, how does he/she evaluate the projects that do not pertain to his/her area, for instance the "implementation of a plan for creating new economic opportunities", which clearly is an economic issue? What happens with socioeconomic projects?

The same applies to people in the infrastructure or transportation fields. Therefore for experts to make comparisons between say seven different alternatives or options, it is necessary to *know them all from the social, environmental and economic point of view.* This does not appear to be very realistic.

b) There is an additional problem. Suppose that we have people with the capacity and ability to examine different issues, but obviously there is no guaranty about the basis or foundation for even one issue. For instance, assume that the *social analyst* is studying projects 1 and 4, that is:

- Construction of storm water drains to stop frequent flooding in the north sector;
- Increase safety by enhancing police surveillance in the east sector.

It is supposed that this analyst, on top of the available technical information will also have input from the people living in the affected sectors. The trouble is, that there are solid grounds for disagreement, *because for the people in the north, flooding is a more serious problem than compared with safety in the east, and vice versa.*

c) Another difficulty is common in some projects.
Assume that there is a project **X** which is compared with project **Y**, and it is found that there is a strong preference for project over project **Y**, and so it gets a 7. Suppose now that there is another project **Z** , which when compared with **X** gets a 9.

Well, nothing wrong with this. We write **7** in the comparison of **X** with **Y**, and a **9** when we compare **Z** with **X**. The hurdle is that, **Y** and **Z** are *both dependent*, that is, the execution of **Z** also implies the execution of **Y**, which obtained a lower rate when we compared **X** and **Y**.

It appears then that this type of problem cannot be solved by this method.

How could we solve it?

Well, instead of comparing alternatives between them regarding a certain criterion, we could utilize another procedure, for example, using a common *weighting system for all projects,* say for instance that we use for importance any number between 1 and 10. "1" corresponding to the lowest importance and "10" to the highest. So, if the appraisal is based on this system, it is possible to ask the social specialist something like this:

How do you rate the construction of storm water drains in the north of the city considering the needs of the people?. Use the 1-10 scale.

He or she, can study this problem, and say: *"Well, because of its importance I assign it an 8"*

Other peoples' opinions can also be gathered about this particular problem, a value assigned and an average computed. The same could be done for the other projects. However, this method cannot be employed here because there is no guarantee of getting an asymmetrical matrix, which as explained, is fundamental in AHP calculations. So, as attractive as it is, AHP --- like any other method - --has its problems, but it does not reduce its efficiency and usefulness.

5.9 Mathematical Programming (MP)
Description

This technique consists in constructing a mathematical model, where real life conditions are represented as closely as possible. Starting with a list of alternatives or options, and *establishing a goal or objective*, the model delivers the *optimal ranking* or *hierarchy between alternatives and in accordance with the objective*. This selection is based on a set of criteria, that is then used to *gauge the contribution of each alternative to the attainment of the objective.* A big advantage of this technique is that it permits *setting up limits to the different criteria, either minimum, maximum or both values simultaneously.*

For instance a certain criterion related to human density in an area could have a maximum limit of 1,073 persons/km^2 and a minimum of 329 persons/km^2. Another criterion could establish a maximum SO_2 limit of 0.025 mg/Nm^3. Notice that both criteria have different units of measure.

All the data *is entered in a spreadsheet*, that is, all the projects or alternatives and their monetary costs, all criteria and their respective limit values, and the specified objective, which can be the *minimization of total costs* (monetary and environmental), or may be the *maximization of scarce resources*, or other objective.

Once the model is set up, a mathematical optimization method is applied to the data. This procedure processes the information and yields a result. This result indicates either:

- The selection of *only one project* out of the initial listing of alternatives or projects considered;
- A *ranking of projects* if there is interest in their relative importance.

In all cases the solution is optimal, in the sense that no better project can be selected or that no improved combination of projects can be found. The mathematical procedure employs different algorithms depending on the type of problem. The origins of these methods go back to the end of World War II, when the American mathematician George Dantzig, developed the so-called *Simplex Algorithm* to solve this type of problem, which in essence is to find a solution when one looks at the *optimization of the objective subject to many constraints.*

The algorithm used in Mathematical Programming depends on the type of problem:

- If the problem is linear, then usually the Simplex algorithm is utilized.

- If the problem is not linear, tree algorithms such as *Branch and Bounds*, are employed.

- For more complex problems the actual trend is to use the *Genetic Algorithm* based in the process Nature uses to improve species.

The user does not have to worry about the type of algorithm to utilize, since some software detects the characteristics of the problem and selects the appropriate algorithm. There are hundreds of computer programs called *codes* to solve this problem, however, in this book the author uses *Solver* which is an add-in of Excel® and other software, and that is fast, accurate and reliable for this type of EIA applications.

As was mentioned, the input of all information is done through a spreadsheet, which constitutes a *database*, allowing the use of hundreds of mathematical functions incorporated in Excel®, or in other advanced software such as Quatro Pro® and Lotus 123®. Because of this the following operations can be rapidly input into the model and without errors:

- Calculation of the regression and correlation analysis of a set of data;
- Calculation of dose-response functions;
- Determination of trends;
- Ancillary calculations that feed the model;
- Etc.

In all cases, changes can be made in the spreadsheet itself or in any of the ancillary sheets containing especial operations as detailed. The result of the change will be *immediately input into the spreadsheet* and as a consequence new results could be obtained. The advantages of this procedure are obvious because of the possibility of analysis of *"What if....?"* scenarios and for exploring the *sensitivity* of the solution (see Appendix, section A.8).

Sensitivity in particular, could be extremely useful in the stage where the solution or solutions reached are discussed with *stakeholders and public opinion,* since new values, suggestions and amendments can be incorporated into the model, and a solution be known in seconds.

Operation

Using Mathematical Programming for EIA has other advantages such as:

- It is possible to indicate *alternatives that must complement each other*. This is the case for instance of two hydro-projects over the same river, one of them upstream and the other downstream, separated perhaps by hundreds of kilometers. In this case both projects *are not exclusive,* for the construction of one of them implies the automatic selection of the other.
- Sometimes there are alternatives that are *mutually exclusive.*
  This could be the case if a decision has to be taken regarding the construction of *either a tunnel* or *a*

bridge over a river, *but not both*. The model can also contemplate this circumstance.

- As mentioned it is possible to have *different objectives for the same data.* The only thing to do is run the model for each objective. However, there is a model called *Goal Programming*--- a part of Mathematical Programming --- where it is possible to indicate, for the same database, *different and even contradictory objectives, such as maximizing economic development but with minimization of environmental costs.*

- Mathematical Programming allows the automatic calculation of very important economic concepts, such as:

  - *Opportunity cost* (see Appendix, section A.6);
  - *Marginal values* (see definition and explanation in Glossary).
  - *Trade-offs* (see Appendix , section A.6).

One wonders why Mathematical Programming produces all of these results. The answer lies in the mathematical properties of matrices (see Appendix, section A.1.3), and the most important fact is that in MP the *problem is not solved on a pair-wise comparison* as in other methods in Multicriteria Analysis, *because all projects and criteria are considered simultaneously, defining a common place for trade-offs and from where the best solution is found* (see Appendix, section A.3).

Of course, in most cases the solution found employing MP is very similar to the solution found for the same problem using other mathematical methods, such as the Analytical Hierarchy Process (AHP) (see section 5.8), but it is believed that *MP offers more possibilities in the incorporation of any type of criteria and because of the additional information obtained from the model.*

There is a very important notion related to MP. Most problems are solved using the *Linear Programming* algorithm developed by Dantzig or utilizing more modern and powerful methods. The Linear Programming model considers simultaneously *two aspects* of the same case through its *direct problem* and the corresponding *dual problem*. In our application for EIA, the solution of the *direct problem* gives *the optimal selection of alternatives or options,* while *the dual problem* supplies *information about the values imputed to criteria, or in other words,* calculates *the significance of each criterion.*

Marginal values for criteria, that is, the determination of how the final solution changes with a unit change in the criterion value, as well as *opportunity cost (see Appendix, section A.6),* i.e., the foregone opportunity when a resource is used for some specific project or option, are also given in Linear Programming.

Finally, it is interesting to mention, that in some EIA *applications shadow prices* (see Appendix, section A.7) should be used instead of market prices. These shadow prices are also an output in Linear Programming, since they correspond to the above mentioned dual values.

5.9.1 Case study --- Construction of an oil pipeline

Background information

This example analyzes four alternative routes for a project consisting of the construction of an oil pipeline, over a distance of about 1,600 kilometers. There is no limit to the number of alternatives to be considered (can be in the hundreds), or total costs, including those for *remediation measures* for each alternative. In this example, see Table 5.32, the alternatives are appraised utilizing a set of 22 criteria, although any number of criteria can be used. As usual, the word *environment* includes all impacts created by the project in any field (social, economic, environment).

Purpose of this analysis

Its purpose is to select the most appropriate alternative route when all economic costs and criteria involving environmental costs are considered.

Procedure

Scoping:

Consists in the determination of the areas or fields that will be used for appraisal of alternatives. In this case the areas are:

- *Environment,* because some of the routes pass nearby sensitive environmental areas, such as a natural park as well as river crossings.

- *Geology,* considering that along the whole length of the pipeline there are different types of soils, as well as a fractured area and the necessity of drilling to install the pipes under rivers.

- *Economics.* From this point of view some of the routes will cross agricultural cultivated land with different type of crops.

- *Social,* because the project will affect the lives of many people, producing *benefits*, such as employment, but also *disturbances* like the increase of crime during the construction period.

- *Risk.*There are environmental risks because of the potential leaks and spills, sabotage risks because of political unrest, and geological risk due to a fractured zone.

The process starts with construction of Table 5.32, the Data Table, with the selection of the criteria and scoring, that is the values that measure the contribution that each alternative makes to the achievement of the objective, and in accordance with one criterion. *Within each area,* alternatives are appraised against several criteria. This data is obtained from expert opinion, surveys, population polls, etc.

Table 5.32 **Data Table**
(From technical data, expert opinion and public consultation)

	ALTERNATIVES				
	A	B	C	D	
Criteria	**Performance matrix**				**Units**
Environment					
Natural forestry crossing	12	1.1	1.3	12	ha
Natural park crossing	5	5	0	12	ha
Impact on groundwater	27	46	24	201	ha
Number of trees to be cut	852	821	280	669	number
Wetlands to be crossed	0	54.8	0	0	ha

Wildlife sanctuary crossing	2.2	2.3	0	0.6	ha
Fauna migration (during construction and use of a permanent road)	5	5	0	3.5	%
Geology					
Terrain slope >15°	0	3.2	5.2	2.9	ha
Terrain slope 8° - 10 °	67	49.5	63.9	5.2	ha
Terrain slope < 8°	102.6	152.9	45.1	129	ha
Water bodies crossings	322	76	653	56	ha
Tunneling	0	0	86.4	0	meters
Salt pans and mud flats to be transversed by the pipeline	12	19	25	32	km
Number of km. with pipeline above ground	0	56	12	0	km
Economics					
Agricultural land crossing	23.8	42.9	110.6	65.5	ha
Social					
Populated area at less than 2 km. from the pipeline route	65.3	102.3	32.1	45.8	ha
Crime during construction	2	2	3	1	%
People opinion against the pipeline	40	50	35	49	%
Risk					
Risk of soil contamination	2	3	2	3	%
Sabotage risk	5	2	5	3	%
Geological risk	0	2	0	0	%
Political risk	20	10	10	18	%

A normalization is now needed in order to have homogeneous figures. To do that, the largest quantity is selected for each row, and then each one in that row is divided by this selected figure. See Table 5.33. Thus, for the first row:

Maximum figure: 12, which is assigned the value 100. Then, for instance for alternative B and criterion *Natural forestry crossing* the new value is:

$$100 \times 1.1 / 12 = 9.16.$$

The next step is to proceed with the Mathematical Programming Table. See Table 5.34. This is the database, that also has incorporated all *the thresholds values corresponding to the different criteria.* They are in the column with the heading of *Criteria limits,* and can be either a sole value or a set of values for each criterion. In this case each value equates the lowest figure in the corresponding row (indicated in bold). Computed figures from calculation are detailed in column labeled *Results from calculation.*
The column labeled *Differences* shows for each row the difference between thresholds and computed figures. The closer these values, the better.

Table 5.33 **Normalization**
(For each criterion each value is related to the largest value)

	ALTERNATIVES (rounded values)			
	A	**B**	**C**	**D**
Criteria				
Environment				
Natural forestry crossing	100	9	11	100
Natural park crossing	42	42	0	100
Impact on groundwater	13	23	9	100
Number of trees to be cut	100	96	33	79
Wetlands to be crossed	0	100	0	0
Wildlife sanctuary crossing	95	100	0	26
Fauna migration (during construction and use of a permanent road)	100	100	0	70
Geology				
Terrain slope > 15°	0	61	100	55
Terrain slope 8 - 10 °	100	74	95	8
Terrain slope < 8°	67	100	29	84
Water bodies crossings	49	12	100	9
Tunneling	0	0	100	0
Salt pans and mud flats to be crossed by the pipeline	38	59	78	100
Number of km with pipeline above ground	0	100	21	0
Economics				
Agricultural land crossing	22	39	100	59

Social				
Populated area at less than 2 km. from the pipeline route	64	100	31	45
Crime during construction	67	67	100	33
People opinion against the pipeline	80	100	70	98
Risk				
Risk of soil contamination	67	100	67	100
Sabotage risk	100	40	100	60
Geological risk	0	100	0	0
Political risk	100	50	50	90

A *decimal weight*, obtained from expert opinion, technical information and from surveys, has been assigned to each criterion as a percentage (see second column in Table 5.34). *Urban as well as sustainable indicators* can also be used, and the *objective is established as to minimize the monetary costs subject to all criteria imposed.*

Rows that denote a difference of zero indicate that the corresponding criterion belongs to the set of criteria that produced the solution of the problem. Other criteria are equally important but they show some sort of slack in their values regarding the established thresholds.

Table 5.34 **Mathematical Programming**
From normalization Table and data Values from Table 5.33 multiplied by the corresponding weight

		ALTERNATIVES						
		A	B	C	D			
Criteria	**Weig-hts**	**Performance matrix**				**Crite-ria limits**	**Results from calcula-tion**	Differ-ences
Environment								
Natural forestry crossing	*0.09*	9	**0.83**	0.98	9	**0.83**	**0.83**	0
Natural park crossing	*0.10*	4.17	4.17	**0**	10	**0**	**4.17**	4.17
Impact on ground water	*0.02*	0.27	0.46	**0.18**	2	**0.18**	**0.46**	0.28
Number of trees to be cut	*0.01*	1.50	1.45	**0.49**	1.18	**0.49**	**1.45**	0.96
Wetlands to be	*0.01*	**0**	1.20	0	0	**0**	**1.20**	1.20

crossed								
Wildlife sanctuary crossing	*0.02*	1.90	2	**0**	0.52	**0**	**2**	2
Fauna migra-ton (during construction and use of a permanent road)	*0.01*	1.30	1.30	**0**	0.91	**0**	**1.30**	1.30
Geology								
Terrain slope > 15°	*0.04*	**0**	2.45	4	2.22	**0**	**2.45**	2.45
Terrain slope 8° - 10°	*0.07*	7	5.17	6.68	**0.54**	**0.54**	**5.17**	4.63
Terrain slope < 8°	*0.05*	3.36	5	**1.47**	4.22	**1.47**	**5**	3.53
Water bodies crossings	*0.02*	0.99	0.23	2	**0.17**	**0.17**	**0.23**	0.06
Tunneling	*0.02*	**0**	0	2	0	**0**	**0**	0
Salt pans and mud flats to be crossed by the pipeline	*0.08*	**3**	4.75	6.25	8	**3**	**4.75**	1.75
Number of km with pipeline above ground	*0.05*	0	5	1.07	**0**	**0**	**5**	5
Economics								
Agricultural land crossing	*0.12*	**2.58**	4.65	12	7.11	**2.58**	**4.65**	2.07
Social								
Populated area at less than 2 km from the pipeline route	*0.03*	1.91	3	**0.94**	1.34	**0.94**	**3**	2.06
Crime during construction	*0.05*	3.33	3.33	5	**1.67**	**1.67**	**3.33**	1.66
People opinion against the pipeline	*0.02*	1.60	2	**1.40**	1.96	**1.40**	**2**	0.60
Risk								
Risk of soil contami-nation	*0.01*	0.67	1	**0.67**	1	**0.67**	**1**	0.33
Sabotage risk	*0.03*	3	**1.2**	3	1.80	**1.20**	**1.20**	0
Geological risk	*0.02*	**0**	2	0	0	**0**	**2**	2
Political risk	*0.12*	12	6	**5**	10.8	**5**	**6**	1
Total of weighted environmental stress values		**58**	**57**	**53**	**64**			
Restriction to only one alternative to be selected		1	1	1	1	**1**	**1**	0

Table 5.35 shows the results of calculation regarding the selection of the different alternatives. The model indicates with a "1" that it has selected alternative B which is *neither the least expensive nor the alternative with the lowest weight, but it is the alternative that best complies with all criteria.*

Compared with the least expensive alternative (D), there is a higher cost of 1,319 - 1,211 = $ 108 million which is almost 9 % more, and that *represents the economic cost to create a lesser damage to the environment.*

Table 5.35 **Alternatives selection**

	ALTERNATIVES			
	A	**B**	**C**	**D**
Total cost for each alternative (million of US$)	**1.389**	**1.319**	**1.425**	**1.211**
Total weighted environmental stress values (from penultimate row in Table 5.34)	**58**	**57**	**53**	**64**
ALTERNATIVE SELECTED	0	**1**	0	0

Comments on results

Cost wise, alternative D should have been chosen, because it has the lowest construction costs. However, this alternative also has the highest environmental cost (environmental stress). As a consequence, and because of the influence of the environmental stress value, the model chooses alternative B, at a *higher cost* but with *less damage to the environment.* Notice that:

- The model allows us (albeit not shown in here), to *identify and rank the criteria* that have a direct relationship with the solution found, and which are independent of the weight established for each criterion. This information comes automatically when solving the MP model because it is the solution of its *dual* (see Appendix, section A.3).
  To better understand this, it is necessary to consider that *all criteria are important to appraise alternatives since it is the comparison of their trade-off values which make the selection.*
  Nevertheless, once the model selects an alternative, it also indicates the criteria that have the closest relationship with this selection. This is shown in Table 5.34 in a column labeled as *Differences* where is depicted the differences between *Results from calculation* and *Criteria thresholds.*

- The model gives the *marginal value (see Glossary)* for each of these criteria, that allows us to know how much *the total cost will change if an additional unit is considered for each one of the selected criterion.*

- For "what if.......?" scenarios and sensitivity analysis, it is only necessary to change the corresponding value of the analyzed criterion. It is possible to change a criterion threshold value at a selected time, or many criteria threshold values simultaneously.

- This analysis was performed with one objective in mind, but it can be used as well for other objectives. This is easily done once the corresponding values for the new objective are known.

Conclusion

As can be appreciated, MP is a very powerful tool to analyze projects within the EIA context, for it not only gives information about the alternatives, or even selecting just one of them, but also provides fundamental intelligence about opportunity costs and shadow prices. It has also the advantage of allowing the use of thresholds, a fact of great importance in the case of analysis especially linked with the *carrying capacity of the environment* (see section 1.11). There is also a technique called Goal Programming, that belongs to the MP family of techniques and that can treat the same type of problems as illustrated here, but also has much more flexibility since it also allows the consideration of different and even conflicting objectives. For all of these reasons it is believed that MP is the tool of the future for EIA analysis.

5.10 Actual examples of multicriteria analysis

Examples

A brief description and comments are now offered on some recent and actual studies in the field of EIA. By noting their Internet addresses the reader can easily access them. No judgment is made on any aspect of these studies. This author's only remarks refer to those particular issues that have been contemplated in each case with the intention that the reader be aware of the scope of situations, different kinds of projects, and *especially the set of criteria used by each author,* and in each example it is pointed out what are considered the most interesting characteristics.

Example 1: Selection of waste incinerators (several countries)

Title: **"Impact assessment and authorization procedure for installations with major environmental risks"**
Author: **A. Rabl *et al.***
Address: www-cenerg.ensmp.fr/rabl/pdf/fullreport.pdf
(see also cited references on the Internet by industries)

This project calls for the selection between different alternatives for courses of action or several different methods for the best way to treat solid wastes. There could be an alternative for a landfill construction, another for burning, another using pyrolisis, etc. In these cases *the selection of a method precludes all others.* This study, published in 1999, extensively analyses the construction of different types of waste incinerators to take place in Spain, Belgium, France, Denmark, Switzerland and the U.K. In this case air emissions are of paramount consideration, and in reality, if the five countries work together there could even be a mix of alternatives, depending on cost considerations, as well as atmospheric conditions.

This report is also a very good example of spatial distribution of health risks, and as stated by the Authors ***"Most of the risk is imposed on the people living beyond a radius of 25 km"......"Decision is made locally whereas costs and benefits are incurred over a much wider region".*** These remarks couldn't be more accurate, and it is for this reason that when analyzing projects in the built-in environment, it is necessary to include in the urban study the concept of *city footprint (see Glossary).*

Example 2: Selection of a sewage system (Norway)

Title: **"The sustainability of conventional versus nature based sewerage systems"**
Authors: **Oddvar Georg Lindholm and Kine Halvorsen Thorèn**
Address:
www.cf.ac.uk/research/bost8/case/watersewerage/conventionalnaturesewerage.html

There are cases when one alternative has to be selected out of a listing of several, but *one precluding* the others. Each one has its pro's and con's, in relation with effects created, costs, etc., and obviously only one of them is to be selected. This is an excellent example of this type of project and analysis. The authors examine two alternatives for a small populated area near Oslo, Norway. Houses in the area are already connected to *cesspools,* and one alternative is the *status quo*, i.e., do nothing, and continue with this sewage system.

The other is to incorporate this area into the Oslo sewage network. Lindholm *et al* analyze this problem considering *thirteen different criteria,* involving the economic, environmental, public health and safety fields, and develop very useful *penalty factors,* which facilitate the selection.

Example 3: Options to alleviate traffic congestion (The Netherlands)

Title: **"Citizen values assessment: incorporating citizen's value judgments in environmental impact assessment"**
Authors: **Stolp A. *et al.*** (see Bibliography --- Stolp A. *et al*, 2002)

In this paper there is a special emphasis on *social impact assessment and public participation*, and as the authors state ***"An adequate and representative consideration of citizen's judgments of environmental attributes, and how project alternatives may affect these qualities, requires a structured research process, which citizen values assessment provides".***

This paper introduces the concept of *citizen values assessment* (CVA), which was developed in The Netherlands. The case study included in this paper, is for a highway option for the city of Rotterdam, and shows a useful analysis of citizen values profile as well as an assessment matrix.

Example 4: Construction of an overhead power line (South Africa)

Title: **"Use of geographic information systems in an environmental impact assessment of an overhead power line"**
Authors: **Warner L.L. *et al*** (see Bibliography - --Warner L.L and R.D. Diab, 2002).

Describes an actual project with the use of GIS for the construction of an overhead power line. The paper analyses the construction of a power line between two points A and C in two legs: From A to B and then from B to C. There are several routes in each segment, which are subject to environmental, land use, slope, cadastral and access criteria. The authors employ GIS and other methods, which allows them to determine the best route for each segment. The interesting thing about this example is the way it considers *effects related to the population.* For instance as criteria, amongst eleven others, they use:

1. Number of rural homesteads in servitude;
2. Number of formal homesteads;
3. Farm dams;
4. Roads crossed;
5. Bends in the route;
6. Etc.

Example 5: Routing for an oil pipeline

Title: **"Inadequate route selection oil pipeline through Georgia"**
Author: **Commission for Environmental Impact Assessment, The Netherlands**
Address: www.eia.nl/english/news/

This example refers to the construction of the Baku - Tibilisi - Ceyhas oil pipeline which encompasses three countries in the Middle East area, but analyses the EIA of the project within the borders of one of them (Republic of Georgia). This study uses seven attributes in the environment, social and safety areas. Very interesting in this analysis is the *consideration of risk.* From this point of view the report considers eight different classes of risk, namely:

1. Terrorism;
2. Sabotage;
3. Military riot;
4. Separation;
5. Civil unrest;
6. External influence;
7. Kidnapping;
8. Criminality.

Example 6: Construction of an LPG pipeline (India)

Title: **"Summary Environmental Impact Assessment LPG Pipeline Project in India"**
Authors: **Engineers India Ltd (EIL), Gas Authority of India LTD (GAIL), National Remote Sensing Agency**
Address: www.adb.org/dcuments/environment/inn/ind-lpg-pipeline.pdf

Related with the construction of an LPG (see Glossary), pipeline between two points distant 1,800 km in the Indian Subcontinent. At the moment of this study, the fluid was transported by rail and trucks. This project, other than the usual criteria about environment, economics and social issues, puts some emphasis on *risk and the environment*. Regarding risk it emphasizes the direct beneficial effects of the project, since LPG transportation by pipeline is by far much safer than rail and/or road transportation, with the ever present possibility of an accident and further consequences. The authors point out the convenience of the pipeline approach when compared with the present ones, considering not only that the pipeline produces much less pollution (including the operation of the pumping and boosting stations), than the emissions from locomotives and trucks. On the other hand, there is an indirect benefit since the consumption of fossil fuels by the conventional means of transportation will decrease significantly.

Example 7: Ranking medium hydro-electric projects (Nepal)

Title **"Medium Hydro-Power Study Project, Nepal"**
Authors: **Nepal Electricity Authority, and Canadian International Water and Energy Consultants**
Address: www.south-asia.com/mhsp/mhsp.htm

The study analyses 16 alternatives for medium hydro-projects in Nepal. In this case the project calls for a *ranking of projects to be recommended, other projects which can be left for future considerations, and finally for projects on which no further action should be taken*. At the time of the study one of the projects was underway, and since another project *was linked* with this first one, because both are in the same hydrological basin, a condition was established that this second project has also to be executed. This is a good example of *complementary projects* but there are also *exclusive projects* in this scheme. The authors consider ten attributes or criteria, incorporating some interesting issues such as projects in a geological fault, risk analysis, and *engineering problems like difficulties in accessing the construction sites and for the erection of the transmission lines.*

Example 8: Policy for energy generation (Canada)

Title: **"Strategic environmental assessment of Canadian energy policy"**
Author: **Noble B.** (see Bibliography - Noble B. M., 2002).

In this strategic environmental assessment five energy scenarios are evaluated to determine the most suitable option for energy generation to be developed to cover the Canadian needs up to the year 2050. The author proposes a methodology in seven phases starting with *scoping the issues* and ending with "*Identifying the best practicable environmental option",* and using the Analytical Hierarchy Process approach (see section 5.8). This report poses five key strategic questions related to the strategy guiding the development of energy policy.

The five options are:

1. Status quo, focusing on energy conservation;
2. Nuclear energy, natural gas and refined petroleum products;
3. Introduction of renewable energies, supplemented with major increases in natural gas supply and refined petroleum products;
4. Phasing out nuclear energy, and increasing use of coal, natural gas and refined petroleum products;
5. Meet demand with natural gas supply and refined petroleum products, supplemented with small increase in coal and hydro-power.

To evaluate these options, Noble utilizes eleven criteria related to emissions, hazardous waste, resource efficiency, heritage preservation, public health and safety and others, and thus covering a wide spectrum of issues. Noble also deals with an aspect that not many people consider: *the impact significance or criteria weighting.*

5.11 Risk Analysis (RA)

Description

Risk has been mentioned many times in this book, however no explanation of the concept has yet been given. It can be defined as the existing possibility of some sort of damage occurring in the future as a consequence of the project's actions. Risk involves both human health and ecosystems alike, and risk analysis works with probabilities of occurrence. In this context it tries to find the distribution of probabilities for variables whose *mean value is not certain.* Three types of risks are considered here:

- *Risk due to human error* during the construction stage of the project. There are always risks that accidents will occur during the construction stage of a project, e.g., fatal falls, people hit by falling objects, electrocution, burns, etc., resulting in injury and sometimes death. For these reasons the project has to adhere to regulations such as those established by OSHA (Occupational Safety and Health Act) and managers must, in all large undertakings, take mandatory courses on work safety.

- *Risk due to natural causes*, like earthquakes, landslides, flooding, etc. Here geological studies can give an idea of the class of risk involved in the project. If there is a geological fracture in the place where a dam is supposed to be built, it is obvious that another location must be chosen. On the other hand, if in a road project there is a risk of rocks falling, it could be possible to consolidate the terrain and build sheds for protection.

- *Risk due to human actions* or inevitable consequences of the effects produced by the project. This book concerns itself mainly with this last type of risk, which presumably would take place in the future, once the project construction is underway. As a consequence it is necessary to involve in the EIA an environmental risk appraisal that includes natural risks.

Risks due to human actions can be linked with uncertainties due to:

- Human behavior;
- Equipment failure;
- Weather;
- Sabotage;
- Political issues;
- Material failure;
- Operation problems;
- Etc.

It would be valuable for a project analyst to have a firm grasp of the risks involved; however, this is a complex issue and specialized publications should be consulted. However, it is useful to point out some risk examples in different activities. See Table 5.36.

Table 5.36 **Associated risks**

Type of projects	**Effects**	**Some examples**
Land use	. Structural risk. . Cave in risks.	- Damage to polders in The Netherlands, when in 1953 more than normal tides broke the dikes and destroyed more than 3,000 houses. - Construction of an industrial plant on a decommissioned landfill.

Petroleun refineries	. Air contamination. . Soil contamination. . Explosions.	- Emission of noxious gases such as benzene, toluene, xylene, as well as suspected carcinogenics. - Leaking of hazardous waste as well as disposal on the soil of some residues from production. - In 1997 an explosion killed four workers at the Tosco refinery in the USA.
Pulp and paper	. Contamination by using chloride compounds for bleaching.	- Pulp and paper plants produce high volumes of wastewater with organochlorine contaminants, which must be treated before their discharge into the waterways.
Steel mills	. Soil contamination. . Air contamination.	- In some plants dust from the emission control equipment has been disposed at the site. - It has been discovered that air pollution from steel mills can produce some genetic damage to rodents.
Chemical industries	. Releases of noxious gases.	- Accident in Bhopal, India (see section 6.1).
Fishing	. Depletion risk.	- Over fishing of the "anchoveta" off the Peruvian coast in the 1950's.
Transporta-tion	. Spills of chemicals. from derailed cars. . Oil spills from ships.	- Sulfuric acid leak from derailed train in Knoxville, USA, in 2002. - Exxon Valdez, which spilled 11 million gallons of crude oil in Alaska in 1989.
Oil and gas pipelines	. Leaks from faulty connections. . Pipe breakdown. . Sabotage.	- Leaking in the pipelines that cross the Alaska's Cook Inlet basin, with an average of a spill a month. (Report realized by Cook Inlet Keeper). - Ecuador pipeline breakdown in 1999. - On June 2003 saboteurs damaged an Iraqi oil pipeline.
Wastes	. Heavy metals leaching.	- Mechanical load and chemical action can provoke breaking of the plastic lining in landfills, allowing heavy metals to contaminate aquifers.
Agriculture	. Too much water consumption. . Too many chemicals in the soil.	- Overused of the Ogallala aquifer, USA. - Cotton project using water from the Amu Darya River. Soil heavily contaminated with pesticides.
Dams	. Break. . Trigger of tremors.	- Break of the Vajont Dan in Northern Italy (see section 2.2.2.13). - There are fears that lakes formed by a dam can trigger tremors.
Pharmaceu-tical industry	. Health impacts.	- Thalidomide drug released in Germany in the late 1950's causing birth defects.

Risk analysis is often based on calculating a main value and determining its probability. However, a more useful measurement is its standard deviation, since the larger this deviation, the greater the uncertainty. So, uncertainty can be viewed as the *level of confidence regarding a specific probability estimate.* If the standard deviation is very small, the uncertainty is also small for there are no large variations expected in the main value. The opposite is also true, since if the standard deviation has a large value, then there is a large uncertainty.

However, the good news is that in many cases uncertainties can be reduced, and the problem consists then in the *determination of the factors causing uncertainty*. Because uncertainty is an old problem in many types of situations, including economic analysis, a *sensitivity analysis* is performed to determine the variations in the outcome produced by changes in some parameters.

For instance in the cost-benefit analysis of a commercial project a Net Present Value (NPV), and/or a Rate of Return (ROR) indicate if the project is economically feasible when changes in certain parameters are considered.

If the project is intended to produce a saleable item such as portland cement for instance, there is no certitude about the costs of the process to manufacture it, no assurance with reference to variations of the sale price per unit, and no certainty regarding the volume to be demanded annually. In most cases one works with a mean value for these parameters, but there is uncertainty about their variations, and a strong enough variation in any of these parameters can render the whole operation unprofitable.

For this reason it is usual to perform what is called *sensitivity analysis* (see Appendix, section A.8), where the parameters are varied individually or jointly, and the outcome, in this case the NPV or the ROR, analyzed. The above mentioned parameters are usually known with a certain degree of confidence since they reflect normal production procedures, as well as known sale prices (either considering similar products existing in the market and/or from a market analysis), and a certain demand according to the expected penetration in the market. It is also viable to work with both the most optimistic and the most pessimistic scenarios for each parameter, or even to compute how much a parameter must change for the venture to have zero profit.

But in other cases this kind of information is nonexistent, e.g., information regarding the effect of a new policy about a proposed tax reduction for clean production. In these circumstances it could be useful to utilize a technique such as the Delphi method (see section 4.3.7).

5.12 Techniques comparison

Table 5.37 shows a comparison of the different techniques. As expected not all the advantages and disadvantages of each process are mentioned. The main purpose of this Table is to establish a comparison between all of them for the reader to have a condensed scenario, and once he/she believe that a particular technique can help in his/her project a more detailed information as well as examples are in the corresponding sections and in the Biblography cited in Internet, as well as in the Bibliography.

Table 5.37 **Techniques comparison**

Technique	Main features	Advantages	Disadvantages
Environmental Impact Assessment	- Provides the main structure and data for all of the other techniques. - Assesses the impact of an activity on the environment.	- Gives a systematic account of the significant environmental effects.	- Does not provide information on cumulative effects.
Cost-Benefit Analysis	- Estimates the social consequences of a project, and works with the willingness to pay for the benefits as well as the willingness to accept a compensation.	- It is easy to use and calculate since it compares the gains in reduction of an environmental damage with the total cost to achieve that reduction.	- Since it works with non-market values sometimes it is necessary to work with shadow prices which are not easy to determine.
Cost-Effectiveness Analysis	- Used when the objective cannot be valuated but can be defined. It seeks to meet the objective at the least cost.	- It is useful because combines economic development with environmental objectives.	- Does not consider secondary effects.
Life Cycle Assessment	- Quantifies the inputs which go into a project in the supply chain, during all of its life.	- The procedure is a good one to identify components and actions that participate in a project. - Independent of price.	- Lots of data needed, as well as a thorough information about the supply chain. - Does not consider some social issues such as the loss of amenities. - It is difficult to define the boundaries.
Input-Output Analysis	- Given a certain project, determines the cumulative inputs it needs, or the cumulative contamination it produces. - Can be used to deliver information on cumulative impacts.	- Considers the whole economy of a country, which implies to take into account the direct and indirect impacts produced by a project.	- Lots of information needed. - Usually shows lack of desegregation, which can produce inaccurate results. - Assumes a linear relationship in the supply chain.
Geographic Information Systems	- Provide spatial information on many different issues and has the ability to superimpose the effects and impacts.	- Very useful to indicate relationships and effects interrelationships.	- Usually an expensive proposition - Does not show temporal effects.

Analytical Hierarchy Process	- Helps the decision making process by providing weight to criteria and alternatives, as well as a ranking of alternatives.	- Provides a good selection of alternatives for the stakeholders to decide. - Easy to understand and employs a dedicated software. - Works well for subjective issues.	- The mathematical basis of the method has been somehow questioned. - On the other hand it is difficult to explain to a non-mathematical audience. - The method is subjective. - Is not too appropriate for a large number of alternatives.
Mathematical Programming	- Helps the decision making process by selecting an alternative, or providing a ranking of them and by determining non-subjective values to criteria. - It supplies an optimal selection of alternatives considering many different criteria and different objectives.	- Allows the determination of criteria weights without human intervention. - Complements LCA and IO analysis. - Can work with a large number of alternatives and criteria.	- It is based in matrix algebra and can be difficult to explain. - There is not a dedicated software for EIA since there are hundreds of codes which can solve the problem. - The formulation of a problem can be difficult.
Risk Assessment	- Estimates the probabilities and degree of vulnerability to humans and the environment.	- Establishes a cause and effect relationship.	-The determination of risk values is not an easy task, in many cases because experiments with lab animals don't replicate too well in human beings. -It is also necessary to know the standard deviation of the mean value.

5.13 Strategic Environmental Assessment

(see definition in section 1.5)

5.13. 1 Case study ---Policy implementation for paper recycling

Background information

This case study deals with the identification of possible options for a policy to promote recycling in the paper industry. (This example has been prepared with source data from the Environment Canada publication entitled *Strategic Environmental Assessment at Environment Canada*, page 15). (See Bibliography - Environment Canada - Strategic Environmental Assessment, 1995).

It is assumed that a governmental body wants to analyze the potential options to promote recycling in the paper industry.

Objective

Minimization of implementation costs.

Alternatives or options

Options are as follows:

1. *Status quo* (existing condition): No policy to encourage government involvement in the recycling industry.
2. *Incentives:* Government might provide grants for tax relief.
3. *Legal measures:* It is possible to pass laws requiring mandatory recycling.
4. *Recognition:* Publicize efforts of good corporate citizens.
5. *Information:* Distribute information about recycling.
6. *Research:* The government undertakes new research to improve recycling technology.
7. *Marketing:* Assist industry in marketing recycled products.

Data on options:

Incentives: It is estimated that a 17 % reduction in environmental costs can be achieved.
Legal measures: It is believed that they can be responsible for 20 % reduction.
Recognition: Can provide perhaps a 5 % reduction.
Information: Gives only a 20 % reduction.
Research: Provides perhaps a 35 % reduction.
Marketing: Can encourage consumption and produce a 20 % reduction.

Table 5.38 depicts the information regarding the implementation costs in million of dollars, as well as the estimate in reduction in environmental costs.

Table 5.38 **Implementation costs**

	1 Status quo	**2 Incen-tives**	**3 Legal meas-ures**	**4 Recog-nition**	**5 Infor-mation**	**6 Re-search**	**7 Mar-keting**
Estimate of implemen-tation costs (millions)	260	40	32	10	5	37	52
Improvement estimate	0.30	0.17	0.20	0.05	0.20	0.35	0.20

Criteria

The following criteria are considered:

1. *Measures of options acceptance:*
   Considers some combination of options, so there are as many rows (criteria) as necessary (in this case 12).

2. *Measures of options feasibility* (100 % feasible equals 1), involving seven feasibility criteria.

3. *Measures of effect of each option* (positive or adverse) and with probability attached (products, activities or events), as follows:

Positive (and corresponding probabilities):

- Contribution to cleaning of water streams;
- Capacity to sell technology;
- Savings in logging, transportation and chipping;
- Good advertisement;
- Savings in legal charges;
- Government doing research;
- Production diversification and new products;
- Government funding in marketing;
- Government money for grants and tax relief.

Adverse (and corresponding probabilities):

- Greater production cost and loss of revenue;

- Involves research costs;
- Involves some capital cost in equipment;
- Involves storing capacity;
- In the long run measures can jeopardize production;
- Expenditures;
- Government research is also used by competition.

4. *Measures of public opinion*

Values given by citizens on the different options as follows:

- 3 = Option not acceptable
- 2 = Option should not proceed
- 1 = Option should be discouraged
1 = Option should be encouraged
2 = Option is a good one but limiting government involvement
3 = Option should address responsibility of paper mills

Explanation of each of these values is in Tables 5.39, 5.40, 5.41a, 5.41b and 5.42.
Thresholds values are indicated for each criterion on the last column at the right.
Using Mathematical Programming to solve this very complex problem, the last row of Table 5.43 shows as a result the percentages values assigned to each alternative or option.

ISSUES RELATED WITH OPTIONS

Table 5.39 **Measures of option acceptance.**

	OPTIONS										
Comments	**1**	**2**	**3**	**4**	**5**	**6**	**7**	**From computation**		**Thresholds**	
Expected acceptance	0.2							0.00	<	0.20	
Expected acceptance	0.2	0.27						0.04	<	0.50	*Status quo and incentives can be linked to improve action*
Expected acceptance	0.2		0.90					0.18	<	0.60	*Linked status quo and legal measures*

Expected acceptance	0.2					0.22		0.02	<	0.40	*Linked status quo and research*
Expected acceptance			0.75			0.22	0.30	0.23	<	0.75	*Linked Legal measures, Research and marketing*
Expected acceptance		0.27						0.04	<	0.27	
Expected acceptance			0.75					0.15	<	0.75	
Expected acceptance				0.9				0.14	<	0.90	
Expected acceptance					0.95			0.21	<	0.95	
Expected acceptance						0.22		0.02	<	0.22	
Expected acceptance						0.22	0.30	0.08	<	0.90	*Linked research and marketing*
Expected acceptance							0.30	0.05	<	0.30	
Available funding from govern-ment (millions of dollars)	1	1	1	1	1	1	1	28	<	37	

Table 5.40 **Measures of option feasibility**

OPTIONS

	1	**2**	**3**	**4**	**5**	**6**	**7**	**From compu-tation**		**Thresh-olds**	
Feasibility	0.2							0.00	>	0.00	*Status quo is considered 20 % feasible, since companies can improve, albeit technical, economic and financial problems might arise*

Each value measures how feasible it is to implement an option. (1 = 100 % feasibility)

Feasibility		1						0.15	>	0.15	*Because of costs involved for the government, this option has a feeble feasibility of 15 %*
Feasibility			1					0.20	>	0.20	*This is a feasible option (20%), however, this number must reflect the consequences of potential closing of plants and loss of jobs*
Feasibility				1				0.15	>	0.15	*Has only a feasibility of 15%*
Feasibility					1			0.22	>	0.02	*Option unfeasible (2%)*
Feasibility						1		0.10	>	0.10	*Research can be very costly, so feasibility is in doubt (10%)*
Feasibility							1	0.18	>	0.18	*This is feasible, however, consideration should be given to the government role in marketing industrial products*

Table 5.41a **Measures of effect of each option**

OPTIONS

	1	2	3	4	5	6	7	From computation		Thresholds	
POSITIVE EFFECTS											**Comments**
-Contribution to cleaning water streams	0.90							0.00	<	0.90	*Offers without a doubt a high positive effect (90%), in getting cleaner water courses in 80 % of the*
-Probability	0.80							0.00	<	0.80	

											cases, and depending on the technology used
- Capacity to sell technology - Probability	1.00 0.30							0.00 0.00	< <	1.00 0.30	*Poses no problems. (100% positive effect) for selling new technology, however, its potential has only a 30 % probability*
Savings in logging, transportation and chipping - Probability				1.00 1.00				0.00 0.00	< <	0.85 0.90	*High (85%) recognition will be acknowledged without a doubt, with a high probability of 90 %*
- Good advertisement - Probability	0.85 0.90							0.00 0.00	< <	0.85 0.90	*It will produce a highly (85%) positive effect in advertisement, and there is a 90 % probability that the system will enhance company image*
- Savings in legal charges - Probability			1.00 1.00					0.20 0.20	< <	0.75 0.95	*The compliance (75%) with legal measures, produces savings in avoiding costly legal procedures and with a probability as high as 95 %.*
- Government doing research. - Probability						1.00 1.00		0.10 0.10	< <	0.70 0.35	*It will have a large effect (70%), however, people can complain about the use of federal funding for industrial uses, and it could get a 35 % chance of approval.*

	1	2	3	4	5	6	7	From computation		Thresholds	Comments
- Production diversification and new products.						1.00	1.00	0.28	<	0.85	*These two options have both a positive effect (about 85%), however, diversification has a low chance (30 %) in both research and marketing.*
- Probability						1.00	1.00	0.28	<	0.30	
- Government funding in marketing							1.00	0.18	<	0.50	*This option will have a positive effect, however, people can complain about the use of federal funding. There is a 20 % probability of getting approved*
- Probability							1.00	0.18	<	0.20	
-Government money for grants and tax relief		1.00						0.15	<	0.95	*Incentives could have a large effect (95%), however, the probability of approving substantial grants and tax relief to industry is low (30%)*
- Probability		1.00						0.15	<	0.30	

Table 5. 41 b **Measures of effect of each option**

OPTIONS

NEGATIVE EFFECTS
Comments

	1	2	3	4	5	6	7	From computation		Thresholds	Comments
- Greater production costs and loss of revenue	0.85							0.00	>	-0.85	*In a large extent (85%) this option will increase costs and hurt sales volume with a probability of 60 %*
- Probability	0.60							0.00	<	0.60	

- Research costs	0.45							0.00	>	-0.45	*There will be research costs (45%), albeit something has already been done. The probability for increasing costs is 50 %.*
- Probability	0.50							0.00	<	0.50	
-Some capital costs in equipment	0.50							0.00	>	-0.50	*There is uncertainty (50%) about how much new capital will be involved. It is believed that capital costs could increase by 20 %*
- Probability	0.20							0.00	<	0.20	
- Storing capacity	0.60							0.00	>	-0.60	*This is largely unknown (60%) because some plants need additional space while others can use available space. Its effect has a 50 % probability*
- Probability	0.50							0.00	<	0.50	
-In the long run legal measures can jeopardize production			1.00					0.20	>	-0.70	*Large production problems (70%), are envisaged. Their likelihooding is about 20 %*
- Probability			1.00					0.20	<	0.20	
-Expenditures can be high		1.00						0.15	>	-0.90	*It is largely agreed that costs will exceed advantages in a figure as high as 90 %. There is a 60 % probability of this fact occurring*
- Probability		1.00						0.15	<	0.60	

-Government research is used by competition						1.00		0.10	>	-1.00	*This is a certainty (100%), however, because different plants produce different products, the chance of this occurring is slim (10%)*
- Probability						1.00		0.00	<	0.10	

Table 5.42 **Measures of public opinion**

	OPTIONS										
	1	**2**	**3**	**4**	**5**	**6**	**7**	**From computation**		**Thresh - olds**	
Value = - 3 Option not acceptable							- 3	-0.54	>	-3.00	*Government involvement in marketing is not acceptable*
Value = - 2 Option should not proceed							- 2	-0.20	>	-2.00	*Government involvement in research is not acceptable*
Value = - 1 Option should be discouraged											*No values for this criteria*
Value = + 1 Option should be encouraged	1		1	1				0.35	<	1.00	*Status quo, legal measures and recog- nition should be encouraged*
Value = + 2 Option is a good one but limiting government involvement		2						0.30	<	2.00	*Incentives should be encouraged but limiting government involvement*
Value = + 3 Option should address responsibility of paper mills					3			0.66	<	3.00	*Information should be a responsibility of paper mills*

Table 5.43 **Results**

1 Status quo	2 Incen- tives	3 Legal meas- ures	4 Recog- nition	5 Infor- mation	6 Re- search	7 Mar- keting
0.00	**0.15**	**0.20**	**0.15**	**0.22**	**0.10**	**0.18**

Therefore its appears that the most important option is Information, followed by Legal Measures and Marketing. It is also possible to get only one solution --that is to get the best option or alternative instead of a ranking of them --- although not always feasible, since it involves solving a much more difficult problem of integer programming, which is done by the software upon request.

Internet references for Chapter 5

CONCEPT: MCA
Title: *Application of Multicriteria analysis to urban land-use planning*
Author: Kesuke Matsuhashi
http://www.iiasa.ac.at/Publications/Documents/IR-97-091.pdf

CONCEPT: MCA
Title: *Multicriteria decision analysis*
Provides technical information about different methods.
http://www-ec.njit.edu/~bartel/LectureNotes/VariousElectre.pdf

CONCEPT: AHP
Title: *Formulation of quality strategy using Analytic Hierarchy Process*
Authors: Wen-Jen Han & Wu-Der Tsay Kaohsiung
Provides detailed information about this method.
http://www.sbaer.uca.edu/Research/1998/WDSI/98wds580.txt

CONCEPT: AHP
Title: *Tradeoff analysis*
Very good explanation with example about the use of the AHP and tradeoffs.
http://www.wam.umd.edu/~gouthami/ense/Chapter_9.htm

CONCEPT: AHP
Title: *The Analytical Hierarchy Process--- A Step-by-Step approach*
Authors: Foster Tom S. & LaCava Gerald
Illustrates the use of this methodology in a very simple way using as an example the selection of a car.
file:///C:/WINDOWS/Temporary%20Internet%20Files/Content.IE5/QRX4XIZF/ahp%5B1%5D.ppt#256,1,The Analytical Hierarchy Process: A Step-by-Step Approach

CONCEPT: AHP
Title: *An application of the Analytical Hierarchy Process to international location decision-making*
Authors: Walailak Atthirawong & Bart MacCarthy
Illustrates the use of this methodology for a site selection problem.
http://www-mmd.eng.cam.ac.uk/cim/imnet/papers2002/Atthirawong.pdf

CONCEPT: AHP
Title: *Strategic implementation of infrastructure priority projects: Case study in Palestine*
Authors: Ziara Mohamed, Khaled Nigim, Adnan Enshassi, Bilal M. Ayyub.
Excellent example of the application of this methodology to nine different projects involving roads, education, health and upgrading.
http://www.ece.uwaterloo.ca/~knigim/JNL/Ziara&Nigim-ASCE-infrastructure.pdf

CONCEPT: GIS and AHP
Title: *Selection of settlement areas using GIS and statistical method (Spatial-AHP)*
Authors: E. Yesilnacar & V. Doyuran-Technical University (METU) Campus in Ankara, Turkey.
This method has been used to select suitable sites for settlements in the Middle East.
http://www.shaping-the-future.de/pdf_www/170_paper.pdf

CONCEPT: LCA-EIO-LP
Title: *Environmental Impact Assessment of multiple product system: Using EIO and LCA in a LP framework*
Authors: Vogstad Klaus, Anders Stromman, Edgard Hertwich.
Excellent explanation of the insertion of a local project "Yearly household demand of paper products", (foreground information) into the national economy (background information), and also offering very good links with related issues. The incorporation of Linear Programming did yield some interesting results especially those related with recycling, where it was possible to determine the optimum recycling level considering the

environmental impact. It also offers comparison when IO analysis is included (the whole economy), and shows how an environmental index varies according to the assumptions.
http://www.stud.ntnu.no/~klausv/publications/Leiden2001.pdf

CONCEPT: Linear Programming
Title: *Student resources*
Explanation of Linear Programming in plain English.
http://www.mhhe.com/economics/maurice7/student/linear.mhtml

CONCEPT: Linear Programming
Very clear and non-technical explanation on Linear Programming. Very good for people who wish to have a comprehension of this tool without the mathematical jargon and with very good examples and graphics.
Explains the dual problem of Linear Programming.
http://www.cs.princeton.edu/~wayne/cs423/lectures/lp-4up.pdf

CONCEPT: Linear Programming
Title: *Linear programming*
A non-technical paper written by the creator of LP.
http://www.cas.mcmaster.ca/~se3c03/journal_papers/lp_dantzig.pdf

CONCEPT: Linear Programming and Analytic Hierarchy Process
Title: *Developing inventory and monitoring programs based on multiple objectives*
Authors: Schmoldt Daniel, David Peterson, David Silsbee.
The paper proposes an example with the integration of LP and AHP models
http://www.srs.fs.usda.gov/pubs/ja/ja_schmoldt015.pdf

CONCEPT: Life Cycle Analysis and Input-Output
Title: *The use of EIO-LCA in assessing National Environmental Policies under the Kyoto Protocol: The Portuguese Economy*
Authors: Nhambiu Jorge, Paulo Ferrão
Excellent explanation of the use of LCA and EIO (Environmental Input Output), together with the Hybrid EIO-LCA (HEIO-LCA).
They present a very good example of the contribution to the Global Warming Potential for the fabrication of glass bottles, and make a comparison between the three methods (LCA, EIO and HEIO-LCA).
http://www.keihanna-plaza.co.jp/ictpi2002/proceedings/14-1.pdf

CONCEPT: Cost-Effectiveness Analysis
Title: *Environmental Impact Assessment Principles and Process*
This publication from the United Nations offers good information about EIA and definitions.
http://www.unescap.org/drpad/vc/orientation/M8_1.htm

CONCEPT: Cost-Effectiveness Analysis --- Rural
Title: *Emission cost-effectiveness calculation procedures*
Author: Greg Gilbert
Offers three practical examples for the calculation of this indicator.
http://www.westernpga.org/files/emiscosteff.pdf

CONCEPT: Cost-Effectiveness Analysis - Railway
Title: *Appendix A: Cost-effectiveness methodology*
San Francisco Municipal Railway
This paper illustrates the calculation of this indicator for three different urban railway alternatives. Very clear explanation.
http://www.sfmuni.com/abm_rpts/srtdamend_a.htm

AREA : Cost-Effectiveness Analysis - --Waste Water
Title: *Cost-Effectiveness Analysis of Effluent Limitation Guidelines and Standards for the Centralized Waste Treatment Industry*
Author: William Wheeler, U.S. Environmental Protection Agency.
Extensive paper (49 pages) dealing with cost-effectiveness options for three subcategories of Centralized Water Treatment, measuring the effect of the regulations in terms of reduction in the pounds of pollutants discharged to surface water. The paper considers 146 pollutants and gives their toxic weighting factor.
http://www.epa.gov/ost/guide/cwt/final/effective.pdf

CONCEPT: GIS
Title: *The potential of a GIS based scoping system: An Israeli proposal and case study*
Authors: Haclay Mordechai , Eran Feitelson, Yerahmiel Doytsher
studyhttp://www.casa.ucl.ac.uk/muki/pdf/gisscoping.pdf

CONCEPT : Life Cycle Analysis
Title: *Impact Assessment, Life Cycle, Management*
Provides good links to Ecological Risk Analysis, Environmental Management Systems, Life Cycle Assessment and to other important subjects.
EnviroInfo
Analysis of the supply chain for car manufacturing in Europe.
http://www.deb.uminho.pt/Fontes/enviroinfo/eia.htm

CONCEPT: Risk
Title: *Recent advances in risk assessments and decision analysis*
Author: G. Makus
http://biosci.usc.edu/documents/bisc102-bakus_EIA_recent.pdf

CONCEPT: Cost-Benefit analysis
Title: *Cost-Benefit analysis of private sector - Environmental investment --- A case study of the Kunda Cement Factory*
Author: Karmokolias Y.
http://netec.mcc.ac.uk/BibEc/data/Papers/fthwobafi30.html

CONCEPT: Hedonic price
Title: *Multi Criteria Analysis: A Manual*
DTLR: Office of the Deputy Prime Minister.
Good comments on monetary-based techniques, such as Financial Analysis, Cost-Effectiveness analysis and Cost-Benefit analysis
http://www.odpm.gov.uk/about/multicriteria/04.htm

CONCEPT: Life Cycle Assessment - Automobile industry
Title: *Environmental Information Systems in corporate engineering: case studies, limits and perspectives*
Authors: Seppelt Ralf & Michael Flake
Paper prepared for a German car manufacturer and using fuzzy expert systems and presents two case studies:

- Magnesium door parts. GIS were used here for integration of Life Cycle Inventory;
- Substitution of thermoplastics by natural fibers for covering applications.

http://enviroinfo.isep.at/UI%20200/Seppelt-Flake_Braunschweig.ell-ath.pdf

CONCEPT : Life Cycle Assessment
Title: *Life Cycle emissions distributions within the economy: Implication of life cycle impact assessment*
Author: Norris Gregory A.
This 13 page paper deals with LCA and the direct determination of the pollutant gases they produce, supplying a limit or threshold about how much one has to go backwards in the supply chain.
http://www.lca-net.com/publ/sra.pdf

CONCEPT: IO model and LCA
Title: *Economic input-output life cycle assessment*

Carnegie Mellon Green Design Initiatiave Input-Output analysis. An extremely useful and highly recommended document. It covers 27 categories from Government to Banking, and main industrial activities. In turn, each category, say for instance "Transportation and Utilities", is broken down in many others, for instance "Railroads and Related Services". For each one of these, results can be obtain regarding:

- Economics;
- Conventional pollutants;
- OSHA Safety/ Employment;
- Water use;
- Global warming gases;
- Fertilizers;
- Energy;
- Ores;
- Hazardous wastes;
- External costs;
- Toxic releases by sector;
- Weighted toxics by sector.

As an example, assume that we want to know the conventional pollutants released by a project related to railroads, introducing as our data the cost of our project in US$. The result will indicate the released contaminants and their weights for this activity and for that amount of investment, as follows:

SO_2 = Sulphur Dioxide
CO = Carbon Monoxide
NO_2 = Nitrogen Dioxide
VOC = Volatile Organic Compounds
Lead = Lead particulate emissions (air)
PM10 = Particulate Matter (less than 10 microns in diameter)

If for instance we choose to know about the economic effect that our investment will have in all sectors of the economy, we choose "Economics" and a Table will display this information. In other words we get the multiplier effect of our project.
http://www.eiolca.net/

CONCEPT: Life Cycle Assessment
Title: *Life cycle analysis and assessment*

This paper makes useful comments on this subject and to conventional applications such as recycling and wastes. It is a good history of the development of this tool.
http://www.gdrc.org/uem/lca/life-cycle.html

CONCEPT: LCA
Title: *LCA Inventory*
PréConsultants.

This paper analyses the first stage in the Life Cycle Analysis, that is the complete inventory of all inputs and outputs of industrial processes. It allows one to download the "Users Manual" for an extensive introduction to LCA.

It is also a Tutorial and a demo.
http://www.pre.nl/life_cycle_assessment/life_cycle_inventory.htm

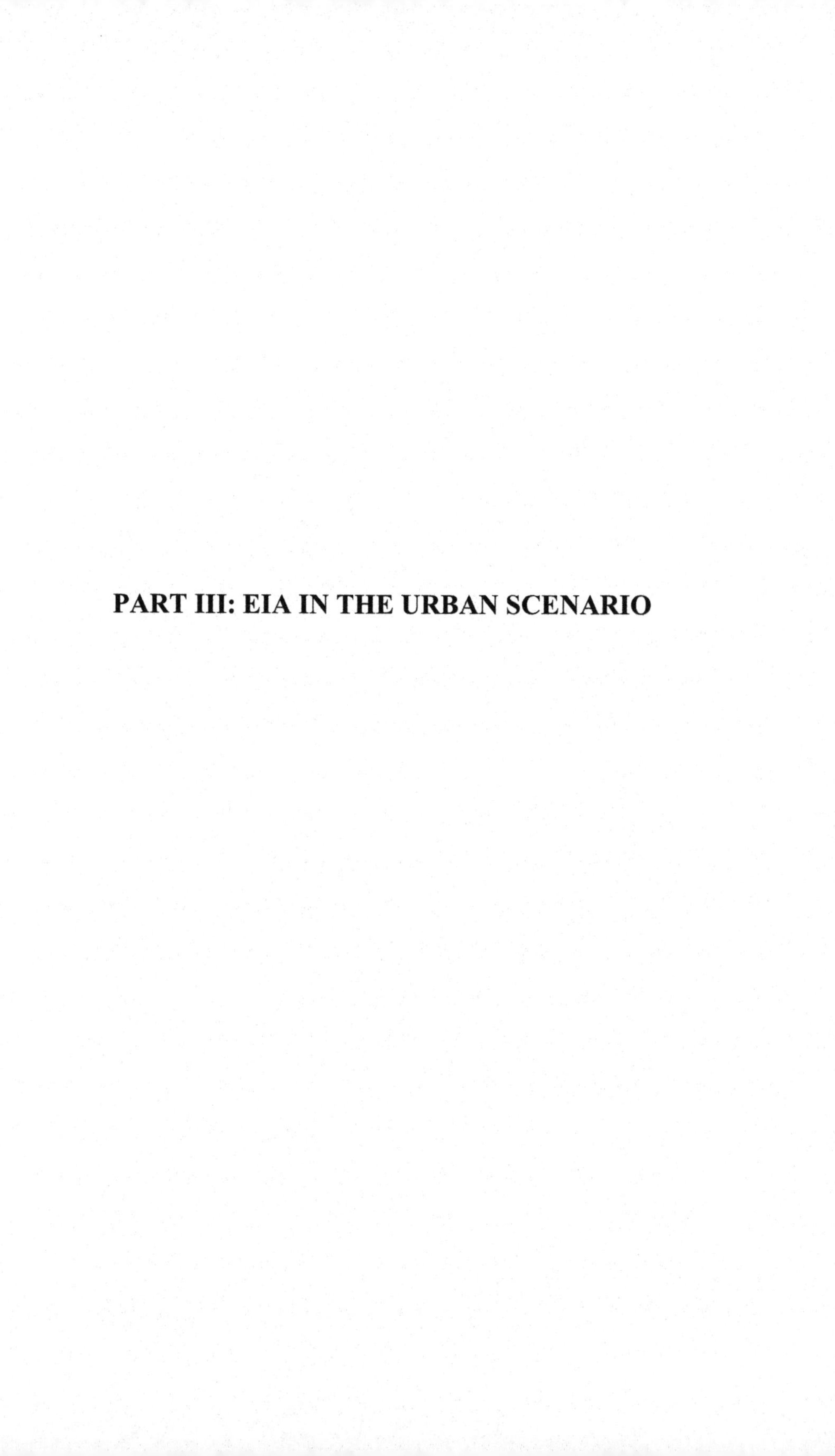

PART III: EIA IN THE URBAN SCENARIO

CHAPTER 6 - EIA AND URBANIZATION

6.1 People's participation

In the past municipalities took measures and executed urban and infrastructure works without consulting the citizens. This is slowly changing as people demand to be heard and even to handle at least a part of the municipal budget. From the EIA point of view it is imperative to inform the people about the potential impacts of a project and also to have their input regarding the way they feel about how the impacts can affect their lives and well being. For that reason *people opinion* is sought and recorded through meetings, polls, surveys, and some sort of a *feedback* is taking place.

Everybody agrees that this is the right path to take, the problem lies in *how to collect information* affecting perhaps thousands of persons, such as in the case of new highway construction. On what issues should people be consulted? From the EIA perspective, people's opinion is very valuable on:

1. *Determination of type of impacts* that can affect the population because of the construction and implementation of a project. For instance a project to install a waste incinerator in an area will produce severe impacts and will for sure generate opposition. Some of the impacts such as excessive truck traffic, noxious emissions, propertydepreciation, etc., are known, but others will come from peoples' input, and this can create a pressure strong enough to kill the project, because *people's perception is quite possibly very different from the perception of analysts and decision-makers.*

2. *How can those impacts affect people?* It is very difficult to convince people that no harm will come from a project, sometimes because there is not enough information about the potential effect on humans. It has happened many times in the past and it can again happen in the future that managers of a project just do not know what impacts will occur. Most probably people don't know either but are suspicious, and human nature tries to protect itself.

3. *What could be the consequences of such impacts?*
   The problem lies in the fact that some consequences are unknown or unpredictable.

   The Bhopal (see Glossary) disaster in India was a consequence of unforeseen circumstances that probably nobody could anticipate. In this case the combination of several factors such as the accidental contact of water with methylisocyanate caused a chemical reaction, that in turn, combined with other chemicals such as chloroform, generated heat to such an extent that the gases formed could not be contained and escaped to the surrounding area, a working-class neighborhood.

It is assumed that more than 3,000 people died and as many as 500,000 people suffered serious injuries. One of the conclusions of the study after the disaster was that it was provoked by a combination of many failures, one of them the lack of adequate training for the people living in the areas or about how to cope with a disaster of this magnitude.

Obviously, this was before that a EIA was a mandatory component of a project and naturally nobody can pretend that, even if the people had been consulted, they could have given a hint of a possible disaster. But what is important to consider is the fact that people *should have been warned* about the dangers of such a project before its construction and during its operation, or could have demanded to be trained and instructed.

At the present time there are different ways to contact people, to make them aware of the impacts of a project and listen to their suggestions and opinions. Because of the *decentralization* of large urban areas there are now in many cities municipal delegations, spread out in order to cover the whole city. On these premises people can get together, be informed by and discuss with City Hall representatives what they think about a project. Generally notes of these meetings are taken or recorded, so there is evidence of what was discussed.

There is also a voting system, usually with computer terminals where people can freely express their thoughts (see section 3.3.6). Of course it is not possible to have a hundred per cent attendance of the whole populated area, and in that sense this voting system should be complemented with surveys, interviews and polls.

6.2 EIA and the city

Fundamentals

It is interesting to note that city planners have used EIA. As an example, in the rehabilitation of neglected areas in many American cities, or in the planning of new communities, there are three basic elements that are being considered:

- Density, that is the number of people per km^2;
- The height of the buildings;
- Transportation.

The idea is to find a compromise between these three subjects.

Regarding *density,* this is a very important aspect since cities, especially in America, tend to expand. This expansion is not new, for it started many years ago with the improvement of transportation, especially with the appearance of the streetcar. It can even be seen today that the main avenues of many cities followed the old streetcar tracks, with communities developing all along them. Nowadays highways are encouraging exodus from the city center to the suburbs. Los Angeles is perhaps the best example of a city where highways are a way of life.

This is a serious problem, for as the city expands it does so at the expense of agricultural land and forest, producing a very strong impact in the ecosystem. On top of that, City Hall needs to expand its transportation network and utilities such as water, sewage trunks, electricity and roads. It is a proven fact that one of the impacts that were not foreseen in the construction of the first highway is that a highway feeds itself, so generally the volume of traffic is always more than the volume expected, and increasing.

Needless to mention the consequences in climate change and in the use of non-renewable resources that this volume of traffic produces are significant, and for this reason the municipal authorities are trying to revitalize the core of the cities and also to increment the density per square kilometer. One way of doing this is promoting the construction of *tall buildings* where the land use can be maximized, but this in turn gives rise to the second subject which is energy consumption, since very tall buildings turn to be expensive and resource consuming.

Another disadvantage of this type of structure is that usually they project a shadow on many other lower buildings depriving their dwellers of illumination and their place under the sun. Besides, there is the risk of fire. Of course, this has nothing to do with the elevation of the buildings, however, it is necessary to recognize that the risk for this danger is increased with the height of the buildings, because there are fewer opportunities for people living on the upper stories to escape in case of fire. The third component is *transportation*, and even if it is true that a high density is essential to have a profitable mass transportation system such as subways, it is also necessary to consider its very elevated capital cost.

As it can be seen, the three discussed elements present a challenge to city planners and EIA specialists alike, for it implies reaching a compromise between different and often conflicting factors. It also means the necessity to consider diverse projects and alternatives involving many different impacts, with some of them very difficult to appraise.

6.2.1 Case study --- Selection of urban intersections

Although this case is not strictly related to EIA it is proposed because it shows how the restrictions imposed, as well as the linking of different alternatives --- which is a common feature in many projects --- can be handled by multicriteria analysis.

Background information

This is an urban project that calls for the construction of eight urban intersections (projects) on four main avenues in the downtown area. These projects are subject to some conditions that are consequences of the urban geography of the city, as well as to some restrictions derived from technical aspects, traffic flow and additional investments.

Detail of intersections. See Figure 6.1

- Intersection 1 on the main avenue as well as intersection 4 on the same thoroughfare;
- Intersection 5 and intersection 6 on another important avenue parallel to the first one;
- Intersection 2;
- Intersection 3, 4 and 6 on the other main avenue intersecting the first two;
- Intersections 6 and 7 on another thoroughfare commencing at intersection 6;
- Intersection 7 and 8 on another avenue.

Conditions which are a consequence of the urban geography. See Figure 6.1

- Execution of intersection 1 implies execution of intersection 4;
- Execution of intersection 1 implies not executing intersection 2;
- In intersection 4 there are two alternatives (1 and 2);
- Execution of intersection 6 implies not executing intersection 4;

If intersection 7 is executed, then intersection 8 must be executed as well.

Execution of intersection 7 implies the construction of a bridge over rail tracks;

Some intersections are linked with others as in 1) and 4) because it is not worth increasing the flow of vehicles in one intersection if in the next intersection they will be at a standstill. Intersection 4) shows an example where the two options are mutually exclusive. This could be the case where it is considered to either build a bridge or a tunnel, but not both.

Whatever the decision in building intersection 4, its execution precludes the construction of intersection 6. Because both are on the same avenue, if intersection 8 is constructed, then it is mandatory to build intersection 7 as well. Rows in Table 6.1 indicate:

-- Row 1: There is a minimum volume of vehicles per hour to be considered in each intersection, with data collected from physical counting over a period of time and projections. The thresholds establish the minimum flow in all intersections, hence the ">" value should be 400 vehicles/hr.

-- Rows 2,3,4,5,6: Specify the conditions derived from the urban geography and discussed above.

-- Row 7: Indicates for each project the necessary land purchase in ($).

-- Row 8: Investment in traffic signals in ($).

-- Row 9: Investments in the construction of storm drains in ($).

-- Row 10: Investment in intersection lighting in ($).

-- Row 11: Investments in embankments and accesses in ($).

-- Row 12: Total investment in additional investments in ($).

-- Row 13: Total investment per intersection in ($).

-- Row 14: Available budget in ($).

Notice that for each of these technical aspects there is a threshold value. It equals the *summation or aggregation* of the values assigned to each intersection for the first criterion. For this reason the ">" sign is used since we want the volume of vehicles that can be handled to be at least as large as the estimated figure of 40,000 vehicles, for the system must not be designed to handle less traffic than estimated.

Other rows show the " = " sign, for we wish to indicate equality in the established conditions. For instance, the second row has a "2" in the threshold column to indicate that there have to be *two* intersections selected, *since execution of intersection "1" implies the execution of intersection "4".*
In the third row there is a "1" in the threshold column because if it is true that intersection "1" is linked with intersection "4", it is only linked with *one* of the alternatives (a bridge or a tunnel) of this intersection. Similar analysis can be done for the other rows.

Notice that rows corresponding to *Additional investments*, have the "<" sign, because we want to be sure that the total expenditures for all intersections and for a certain criterion, be, as a maximum, equal to the available funding.

Now the problem can be solved by MP. This is quite a complicated program, but its purpose is to demonstrate that there could be mutually exclusive alternatives, together with some complementary alternatives and mixed with others that stand alone.

For each of these alternatives and their associated restrictions the EIA has to be performed, and then the intersection/s *selected* which offer the best combination of costs, and minimum environmental and social damage. See Table 6.1; the result indicates the following:

Intersections 1, 2 , 3 , 5, 7 and 8 are selected.
In intersection 4 option number "1" is selected.
Intersection 6 has not been selected.
The result also shows that all restrictions have been met.

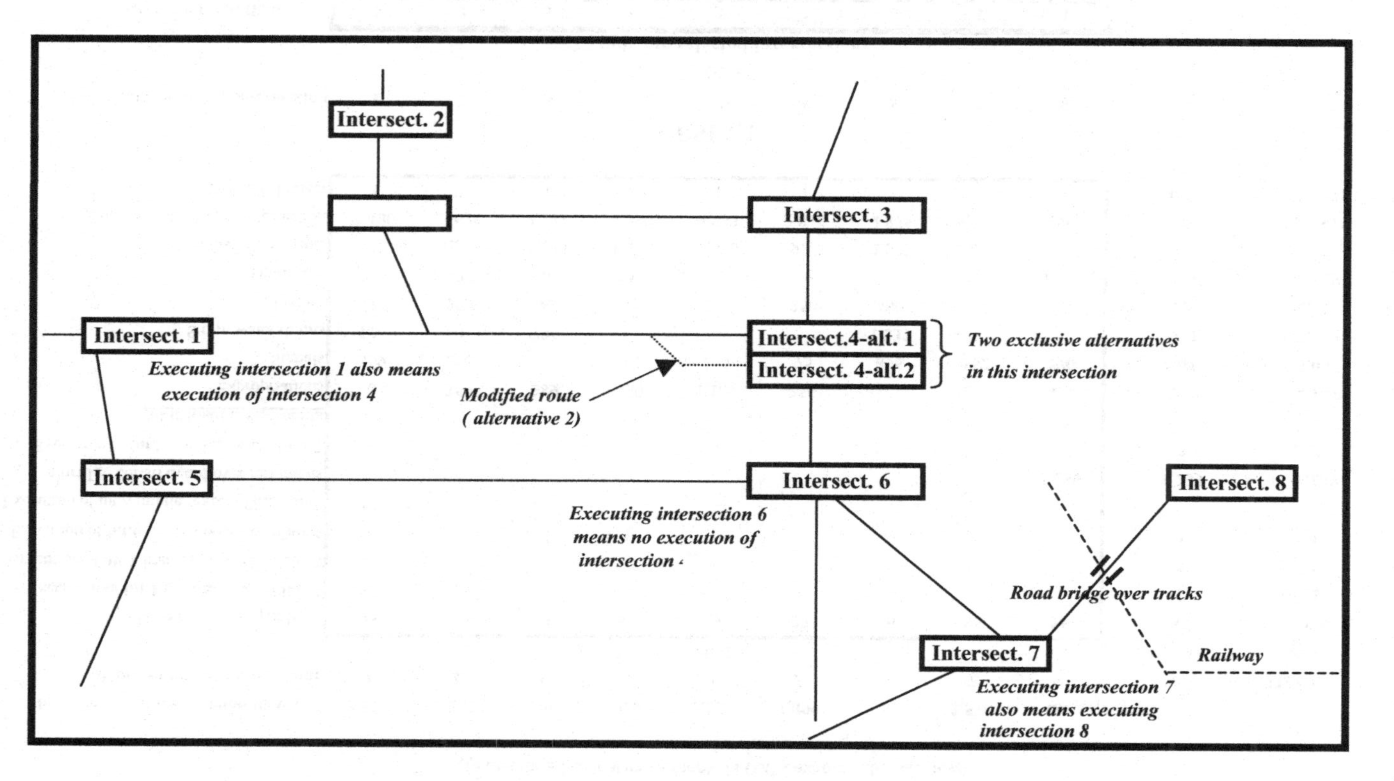

Figure 6.1. **Alternative intersections scheme**

Table 6.1

Selection of urban intersections

(values for additional investments in 000's except in the last row)

Investment in construction [000's $]	**3,570**	**2,257**	**1,096**	**3,046**	**3,257**	**1,890**	**282**	**2,570**	**459**			
Name of urban intersection:	**1**	**2**	**3**	**4**	**4**	**5**	**6**	**7**	**8**	**Comp.**		**Thresh-**
				Alt. 1	**Alt. 2**					**results**		**olds**
Volume of vehicles per hour	53	40	43	89	99	27	76	85	56	**400**	>	**400**
Execution of **int.1** implies exec. of **int. 4**	1			1	1					**2**	=	**2**
Execution of **alt.1** implies no exec. of **alt. 2**				1	1					**1**	=	**1**
Execution of **int.6** implies no exec. of **int.4**				1	1					**1**	=	**1**
Execution of int.8 implies exec. of int.7 and **a bridge constructed** over rail tracks									1,239	**1,239**	=	**1,238**
Execution of **int.8** implies exec. of **int.7**								1	1	**2**	=	**2**
Additional investments												
Expropriations	0	359	289	450	1,100	256		1,237	457	**3,578**	<	**4,147**
Signals	359	256		102	246	75	896	562	570	**2,042**	<	**3,066**
Storm water drains	126	0	796				1,237			**922**	<	**2,158**
Lighting	426	229	90	242	260	250	460	237		**1,487**	<	**2,193**
Embankments		527	0	450	70	346			1,237	**2,249**	<	**2,629**
Total additional investments	910	1,371	1,175	1,244	1,676	927	2,593	2,036	3,503			
Total investment per intersection	4,480	3,628	2,271	4,290	4,933	2,816	2,875	4,606	3,962			**33,862**
Available budget	1	1	1	1	1	1	1	1	1	**26,600**	<	**33,862**

RESULT

Name of urban intersection	**1**	**2**	**3**	**4**	**4**	**5**	**6**	**7**	**8**
				Alt. 1	**Alt. 2**				
				"1" indicate selected intersections					
Selected intersections	1	1	1	0	1	1	0	1	1

6.3 Urban and regional projects

It is necessary to realize the importance of performing EIA for urban projects because of the impacts on people and their health, but also given that in a city the concentration or density of population is much more relevant than in projects that take place in isolated areas. As a consequence the effects of projects whatever their nature, have a much larger impact in an urban nucleus than in the countryside. There are many projects requiring EIA located in cities and in their metropolitan areas. Such projects refer for instance to the construction of:

1. Different kind of manufacturing plants (automobiles, portland cement, food, pharmaceuticals, power stations, etc.);
2. Chemical and petrochemical complexes;
3. Potable water and sewage treatment plants;
4. Landfills;
5. Urban complexes;
6. Industrial parks;
7. Harbors facilities;
8. Roads and highways;
9. Hydro-electric and irrigation dams;
10. Etc.

In all cases there will be a change in the receptors, that is the people and the environment, and for that reason an EIA is required. What is important to consider is that a project, even if it is done in the middle of the city, in reality has impacts that are *spatially felt* in the *metropolitan area* and *beyond.*
Think for instance of an urban sewage project which will collect domestic sewage of an area of the city with 40,000 households, that until that time discharge their waste water into individual cesspools.

It will have of course, a beneficial effect on the environment since the cesspools are contaminating underground water sources. However, it is also necessary to evaluate the impact on the environment of discharging about 375,000 m^3 of treated sewage into a small river traversing the city. Considering that the city is "importing" raw water from a place 156 km. away, and discharging sewage in a place located 17 km. away, it can be seen very clearly that this very urban project (connecting an area of the city to its sewage network), has effects well beyond the urban limits.

But, it is not only water that the city purchases outside its limits. It also procures produce (from its green belt), fruit, grains, meat, construction materials, etc. For this reason it makes sense to consider not only the city but also an area loosely defined, which supplies needed goods, and an area which is the receptor of city refuse, mainly garbage and waste water.

This area has been called by some authors the *footprint of the city* and is usually not related to what is called the *hinterland* which is defined by economic considerations. Naturally, the city also gives employment, produces goods and services, entertainment, education, etc, which benefit its footprint, so there is a symbiosis between these two entities.

Therefore, when analyzing certain urban projects attention has to be paid to their spatial context, and consequently, their impacts analyzed considering this circumstance.

6.4 The execution of joint projects

Usually a large metropolitan area includes many small towns which are satellite cities to the main urban agglomeration. Generally, each one of these small peripheral towns have their own City Hall as well as their own budget. In many cases this budget is supported by taxes collected in a particular town, plus a certain contribution from the provincial or state government, and by federal funding.

There are different ways in which this budget is formed, but one component is always present: *population size.* Some cities have a large number of satellite cities, perhaps more than 10, so a problem arises when there are infrastructure projects that affects some of them. Sewage is one; others are for instance the construction of a highway connecting three or four areas of the metropolitan area. It is obvious that the continuation of a road cannot be halted because one of the towns opposes its construction, since the highway would come to a dead end and would lose a large part of its value. This illustrates the importance of the political input.

In many cities the construction of certain parts of an inter-municipal road may involve the excavation of tunnels under a river or through a mountain, which of course are very expensive undertakings, that some towns are in no condition or willingness to afford. In most of the cases in large cities, this problem is solved because there is a Council which handles all these inter-municipal affairs. It is then up to the Council to collect information regarding the impacts that a certain project will cause within each jurisdiction.

The problem can be very complicated because there could also be a need to pay compensation to areas that have been selected as dumping sites for the city's solid waste.

The progress of a city is usually measured through a set of urban indicators involving different fields such as environment, social, economic, etc. Therefore, it is a good idea to include in the EIA study some of these indicators with their corresponding thresholds, for this will guarantee that the impacts of the project under study will not exceed these limits. This can be done using Multicriteria Analysis and most especially Mathematical Programming.

6.5 Sustainable impact

Another important aspect to consider but which unfortunately is not taken into account very often, is the sustainability of a resource. To illustrate this subject we will use a case study, that corresponds to the construction of houses for low income people.

6.5.1 Case study --- Plan of new dwellings for low-income people

Background information

Assume that the selected area for the relocation of these people is in the south of the city. The project is *economically feasible,* since the undertaking will be paid for by the dwellers with individual loans repayable in up to 25 years and with special financial conditions. So, a cost-benefit analysis for this project will be positive. It is *socially feasible* considering that low income people will have access to decent dwellings, and also that they will be incorporated into the city fabric instead of being marginalized.

From the *environmental point of view there are no objections*, quite the opposite. The slums in this case are at the present time health and environmental time bombs, because added to the lack of water and sanitary services, garbage is thrown on the streets where children share the playground with rodents. The slum area is in a hilly part of the city, so on top of the mentioned problems, there is always the permanent risk of slides during heavy rains.

This is a natural phenomenon, however it has been greatly helped by erosion as a consequence of the logging of a natural forest that existed in the area, where trees were cut by the slum dwellers to be used as fuel for heating and cooking. The area chosen for the new settlement is a plain area, without any remarkable geographic faults, and that will be easily connected to the rest of the city.

Once the people are moved, the slum area will be converted into a park, with thousands of trees to be planted, which will reduce the risk of slides and will provide a recreational area which is badly needed in that part of the city. Water for the new settlement should come from the city water network, however this is not possible because the system is already working at its greatest capacity, and in reality this is one of the most pressing problems in the city. Its inhabitants, at the present time do not have all the water in quantity and quality according to international standards.

The new settlement will be linked to the city sewage network, which also offers some problems, since the treatment plant is already working at its full capacity. Finally, technical studies were conducted to assure the supply of potable water to the new dwellers, and from this point of view the provision will be assured by drilling water wells to reach an aquifer located at a depth of 42 meters. Therefore, everything appears to be ready for the commencement of this project. However, there is one factor that has not been considered: *The supply of potable water is assured, but for how long?*

The question arises because even though these wells will serve the new settlement, the *extracted rate of water per day will exceed the daily rate of water recharged by natural causes.* Considering the population size of the new settlement it is believed that the aquifer feeding these wells will be exhausted after 23 years. Therefore this is a secondary or indirect effect of our project. As a consequence, one impact of this housing project will be the *depletion of a natural resource*, which means that the *carrying capacity of the environment* has not been considered during the technical stage of the project.

A solution has to be found. An alternative is to reduce the population of this settlement in such a way as to have a daily consumption below the threshold established for the aquifer, and because there are other parts of the city that also have some aquifers, it would be possible to relocate there the balance of the dwellers. There is also the problem of waste water treatment since as mentioned the capacity of the actual system is near collapse.

A solution --- albeit not the best --- could be for instance to construct a local cesspool serving the new settlement, or to build, which is much better, a local treatment plant, discharging treated water into a creek that voids into the local river. This case study shows how the incorporation of a criteria threshold in the multicriteria study *could detect a problem that for some reason was not considered in the technical studies.* This example aims to show the importance of *sustainability indicators* which must be added to economic, social, and sustainable criteria

Needless to say, sustainability indicators do not only apply to water, for they can be used for air, soil, forests, land use, etc. As a consequence, when there is a portfolio of projects or alternatives, it could very well be that some of them are not chosen because they violate a sustainable indicator.
Therefore, by examining restrictions or constraints imposed by thresholds in sustainable criteria, it can be detected *additional work that needs to be executed,* and which was not foreseen in the original technical study. In this case the extra work would be the construction of the local water treatment plant for both new settlements.

6.6 Indicators

Indicators have been mentioned several times in this book. But what are they?

Indicators are *measures of a certain activity* and they are called *metric* when they have a cardinal value attached. There are hundreds of indicators involving the most diverse fields, but for the EIA study in urban issues, the interest is mostly centered in urban indicators. A publication such as "Urban Studies" is recommended as a source of quantitive data for urban indicators. (see Bibliography---Flood J.).

One of the best sources is United Nations - Habitat, which you can access through http://www.unchs.org/programmes/guo/. Here there is also information about the newest developments such as The Global Urban Observatory (GUO) and the Local Urban Observatory (LUO) (see Glossary for definitions), as well as a wealth of statistical data on "Human Settlements" and "Citybase Database" at city level.

Table 6.2 has been reproduced from the source mentioned below. A whole and comprehensive definition, significance, methodology, etc. of each indicator and their linkages, are detailed in the mentioned Internet address.

There is also a good guide for municipal indicators produced by Environment Canada, in a publication entitled "State of the Environment Reporting - Municipal State of the Environment Reporting in Canada: Current Status and Future Needs" - Occasional Papers series No. 6, page 33, 1995. The listing is reproduced in Table 6.3.

These indicators are extremely useful in monitoring sustainability, as well as changes and trends in the urban environment. However, in most of the cases, on top of these indicators it is necessary to utilize *local indicators related with local conditions and specific for a city.*

Indicators are also used to compare how good or how bad is the state of the city when compared with other urban agglomerations. To do that one has to compare local values with international standards. However, this could be misleading since it is assumed that all cities have the same problems, and this is far from true.

For instance most cities in the world have problems in the water supply, however this is not the case in many other cities where this vital element is abundant, but obviously, these cities in turn have problems that other cities don't. For instance it wouldn't make sense in a Central American city to have an indicator about the percentage of streets in the city free of snow five hours after a heavy snowfall, but it makes a lot of sense in many cities in North America and northern Europe.

Table 6.2 **Urban indicators**

CHAPTER 1: Shelter	
1. Provide security of tenure	Indicator 1: Tenure types Indicator 2: Evictions
2. Promote the right to adequate housing	Qualitative data: Housing rights Indicator 3: Housing price-to-income ratio
3. Provide equal access to land	Indicator 4: Land price-to-income ratio
4. Promote equal access to credit	Indicator 5: Mortgage and non-mortgage
5. Promote access to basic services	Indicator 6: Access to water Indicator 7: Household connections
CHAPTER 2: Social development and eradication of poverty	
6. Provide equal opportunities for a safe and healthy life	Indicator 8: Under-five mortality Indicator 9: Crime rates Qualitative 2: Urban violence
7. Promote social integration and support disadvantaged groups	Indicator 10: Poor households
8. Promote gender equality in human settlements	Indicator 11: Female-male gaps
CHAPTER 3: Environmental management	
9. Promote geographically-balanced settlement structure	Indicator 12: Urban population growth
10. Manage supply and demand of water in an effective manner	Indicator 13: Water consumption Indicator 14: Price of water

11. Reduce urban pollution	Indicator 15: Air pollution Indicator 16: Wastewater treated Indicator 17: Solid waste disposal
12. Prevent disasters and rebuild settlements	Qualitative data 3: Disaster prevention and mitigation
13. Promote effective and environmentally sound transportation system	Indicator 18: Travel time Indicator 19: Transport models
14. Support mechanisms to prepare and implement local environmental plans and local Agenda 21 initiatives	Qualitative data 4: Local environmental plan
CHAPTER 4: Economic Development	
15. Strengthen small and micro-enterprises, particularly those developed by women	Indicator 20: Informal employment
16. Encourage public-service partnership and stimulate productive employment opportunities	Qualitative 5: Public-private partnerships Indicator 21: City product Indicator 22: Unemployment
CHAPTER 5: Governance	
17. Promote decentralization and strengthen local authorities. 18. Encourage and support participation and civic engagement. 19. Ensure transparent, accountable and efficient governance of towns, cities and metropolitan areas	Qualitative data 6: Level of decentralization Qualitative data 7: Citizen involvement in major planning decisions Qualitative data 8: Transparency and accountability Indicator 23: Local government cooperation and partnerships
CHAPTER 6: International Cooperation	
20. Enhance international cooperation and partnerships	Qualitative data 9: Engagement in international cooperation

Source: United Nations - Urban Indicators Tool Kit - GUIDE
www.unchs.org/programmes/guo/guo_guide.asp
Table reprinted with kind permission of United Nations Habitat.

Table 6.3 **Most common municipal indicators**

Type	Indicator
Biophysical	
	Air pollutants levels (SO_2, TSP, CO, NO_2, O_3, Lead)
	Land area by land use
	Farm woodlots
	Area of wetlands
	Environmentally sensitive areas
	Number of farms
	General pesticide use

	Sand & gravel removal
	Car trips
	Transit readership
	Air quality exceedances
	Bacterial contamination of water
	Phosphorus in water
	Fecal coliform in water
	Nitrogen/nitrate in water
	Discharge of treated sewage
	Destination of processed sewage sludge
	Water consumption
	Water treatment plant connections
	Chlorine in drinking water
	Solid waste generation
	Industrial & commercial waste generation
	Municipal waste generation
	Amount waste land filled
	Municipal landfills
	Recyclables diverted from landfill
	Recycling programs
	Hazardous & special wastes
	Spills response & prevention
	Amount of household hazardous waste collected
	Household hazardous waste sessions
Social	
	Population & growth
	Urban-rural mix
	Employment by sector
	Dwellings units
Health	
	Traffic accidents

Source: Environment Canada
Reprinted with kind permission of Environment Canada

6.7 Indicators selection

It has been seen that there are likely to be several hundreds of indicators pertaining to different areas of the built environment; however it is obvious that it is impossible to handle this number of indicators, let alone to make a comparison between them, especially when they are subject to many different criteria. In other words, if there is an initial listing of a pre-selected set of say 135 indicators, and only 25 to 30 indicators are needed, *how is the selection made?* To answer this question it is necessary to determine what is the *objective* in having that final short list. This objective consists in getting a list including the most important indicators and *condensing the maximum quantity of information.*

6.8 Measuring the quantity of information provided by indicators

How can the quantity of information be measured? Until now it has been assumed that all indicators in the original set provide the same amount of information. But in general it is not so, as a consequence, it would make sense to assign to each indicator a value that will show the quantity of information it can furnish. This value or weight can be determined by analyzing for each indicator its links with other indicators. The stronger the link the larger the value. As an example, we can consider the indicator that shows *water consumption per capita.* This indicator allows us:

- To compute the total volume of water the city requires;
- To make a direct comparison between the rate of water extracted from wells and the recharging rate;
- To calculate the required capacity for the water treatment plant, as well as future demand, based on population growth. This, in turn, is related with the searching for urban and/or suburban space to build the water treatment plant;
- To determine if the population is receiving an adequate quantity of water in accordance with international standards;
- To find out if there is water savings as a consequence of environmental campaigns, measuring the difference between past and present day water consumption;
- To compute the volume of wastewater to be generated;
- To calculate the necessary capacity of the wastewater treatment plant. This, in turn, is related to the searching of urban and/or suburban space to build or to expand the wastewater treatment plant.

Another indicator, for instance one measuring the *building permits* in a year and per 1000 inhabitants, allows us:

- To determine the rate of land use for urban use;
- The loss of agricultural land surrounding the city and the influence exerted by the city on neighboring areas (metropolitan region);
- To learn about the employment to be created in the region, which in turn is an activity with a strong multiplier (see Glossary);
- To determine available space per capita in dwellings, and comparison with international standards.

An indicator such as *crime rate* allows us:

- To infer economic conditions in the city;
- To determine the need to diversify the urban economy;
- To learn about quality of police control in the streets;
- To estimate the influence of unemployment;
- To determine the effects of drug use.

It can be seen that with this type of analysis it is *possible to determine, subjectively, a weight for each indicator relative to the others.* To select indicators out of an initial set, multicriteria analysis can be used. Here there will be in columns the complete set of indicators and in rows the criteria used to select them. If one so wishes, a pre-determined weight can be assigned to each indicator, and derived from the above analysis.

This author has applied Information Theory (IT) (see Glossary), to determine numerically the quantity of information given by a selected set of indicators. This approach relies on the concept of entropy as defined earlier (see section 3.1.10). It is assumed that as an objective a final set of indicators *should provide the maximum amount of information.* Consequently each indicator of the initial set, perhaps hundreds of them, is given a weight which would represent its importance. This importance could be measured by the number of links of such an indicator with others, backwards and forwards. In this way the MP problem will optimize the amount of information provided and subject to all restrictions and criteria imposed, and in so doing will select the *final set of indicators that satisfy all constraints.*

6.8.1 Case study --- Selection of urban sustainable indicators

This example illustrates the use of Multicriteria Analysis (MCA) for the selection of urban sustainable indicators. The list of indicators, sustainable goals and general selection criteria used in this example has been taken from "Developing Indicators of Urban Sustainability: A Focus on the Canadian Experience" - CMCH and Environment Canada. (see Bibliography - MacLaren V.).

Background information

A city wants to develop its own set of urban indicators to be able to monitor progress on actions taken. Usually, out of the hundreds of indicators available from different sources, such as the United Nations - Habitat, the World Bank, some municipalities, etc, it is necessary for practical reasons to choose *only a small set of them.* Otherwise, it would be impossible to deal with --- let alone understand --- the meaning and the relationship of hundreds of indicators. For this example it is assumed that said initial list contains only sixteen indicators, and that only eight of them are to be selected. The sixteen indicators are described in Table 6.4.

Table 6.4 **Set of indicators**

Areas	#	Possible indicators are	Meaning and use
Environmental	1	Air quality	Number of times when values of air quality exceed quality goals
	2	Primary commuting modes	A large quantity of cars produce pollution
	3	Residential water consumption	Useful to monitor water saving efforts
Social	4	Adult literacy rate	It is a social indicator, since it allows for the population to know social programs and to have access to better job opportunities
	5	Low birth weight infants	Low weight increases death risk, and during childhood the danger of neurological and respiratory problems
	6	Crime rate	It is a public safety and social conditions indicator
Economics	7	Employment concentration	Measures economic diversity as a percentage of the labor force employed by the top ten local employers
	8	Building permits	It is a viability indicator in the community, and more specifically of employment opportunities in the construction industry
	9	Unemployment rate	It is an indicator of social stress
Environmental - Social	10	Environmental restoration activities	It refers to the number of people or the percentage of the population participating in volunteer environmental

			restoration activities
	11	Green spaces	A social well-being indicator
Environmental - Economics	12	Defensive expenditures	Measures expenditures made to prevent or compensate for environmental degradation
	13	Environmental elasticity	Compares how countries perform on critical environmental indicators relative to critical economic indicators
Social - Economics	14	Low-incomehouseholds	In is a measure of social stress
	15	Health-care expenditures	It is a measure of management and actions taken by the municipality in behalf of the city population
Environmental -Social -Economics	16	Appropriated carrying capacity	The maximum population that a given region can support in perpetuity

Set of conditions

Indicators to be selected must comply with conditions or specifications regarding:

- Sustainable goals. Table 6.5.
- General selection criteria. Table 6.6.
- Approach. Table 6.7.
- Areas participation. Table 6.8.

Table 6.5 **Sustainable goals**

Number	**Goals**	**These are the objectives we want the indicators to represent**
1	Inter-generational equity	Involves the concept that needs of future generations are as important as needs at the present (social equity) It means a fair share of benefits and costs of existing natural resources.
2	Intra-generational equity	Economic development and a better quality of life have to be improved, but not at the expense of the environment (geographical equity). It means the standard of life has to be increased, but not at the expense of the degradation of neighboring regions.
3	Minimal impact on the environment	It means that any type of releases should not exceed the regenerative capacity of the natural system. This involves for instance monitoring the assimilative capacity of surface water and dispersion capacity of the air.

4	Living off the interest of renewable resources	If a resource is renewable, the population can live off its "interest", but without depleting the source, the "capital". A good example is water from an aquifer, which should be pumped out at a lower rate than its recharging rate. Same for logging activities from forests.
5	Minimal use of non-renewable resources	This is the case of fossil fuel. As a consequence, it is necessary to encourage substitution and recycling.
6	Long-term economic development	Assuring economic vitality is an essential component of urban sustainability. Therefore, it is important to take into account economic, employment and employment concentration factors.
7	Diversity	It is the ability to adapt to changes. From this point of view it is necessary to have a population with a broad and varied technical base.
8	Individual well-being	Individual well-being extends to physical, social and mental aspects. Therefore, it is critical to develop human potential, and basic, cultural and recreational needs for the population.

Table 6.6 **General selection criteria**

Identifi-cation	**Criteria**	**These are the conditions we want the indicators to consider**
B	Representative	It refers to a specified subject or to a broad scope of environmental, social and economic conditions.
F	Understandable by potential users	The scientific content should be understood by people at different levels.
G	Comparable to thresholds or targets	Indicators should be able to be compared with standards either local, provincial or national. This way it will be possible to determine how close or how far we are from something considered "optimum" or "wishable".
H	Comparable with other indicators	In some cases mainly in metropolitan areas it could be interesting to compare progress reached in a determined subject or area.
I	Cost effective	This is an obvious consideration. The indicator must provide information at a reasonable cost.
J	Unambiguous	The indicator should be very precise and unambiguous.
K	Attractive to the media	The indicator must be attractive to be considered by the media, since diffusion is essential.

Table 6.7 **Approach**

Identi-fication	Approaches	These are the aspects of the framework we want the indicators to represent
C	Condition	Indicators that show what is happening, for case the contamination percentage in a river measured by the Biological Demand of Oxygen (BDO_5) (see Glossary).
S	Stress	Indicators that show a problem or state, for instance the discharge of untreated sewage to cesspools.
R	Response	Indicators showing the response to these problems from the City Hall, for example the percentage of treated sewage discharged into a river.

Table 6.8 **Areas participation**

Identi-fication	Areas	These are the different fields in the urban built environment we want the indicators to address
Environ-ment	Environment	Establishes that the result must contain indicators related with environment.
Social	Social	Establishes that the result must contain indicators related with social issues.
Economics	Economic	Establishes that the result must contain indicators related with economic issues.
Env-social	Environmental - Social	Establishes that the result must contain indicators related with environmental and social issues.
Env. - Economics	Environmental - Economic	Establishes that the result must contain indicators related with environmental and economic issues.
Eco.-social	Socialeconomic	Establishes that the result must contain indicators related with social and economic issues.
Env.-Eco-Soc.	Environmental - Socialeconomic	Establishes that the result must contain indicators related with environmental, social and economic issues.

First Step

We first need to construct a matrix containing all the available information. In columns are the indicators, while rows (attributes), depict sustainable goals, general selection criteria, approach, and areas of participation. If a relationship is considered that exists between an *indicator* (in a column), and a *criterion* (in a row), meaning that the *respective criterion is deemed appropriate for measuring a particular indicator*, then the relevant intersection is marked with a "1". For instance, comparing Table 6.4 with Tables 6.5 to 6.8, it is understood that indicator number 6 (*Crime rate),* is related to the following criteria (see Table 6.9).

Table 6.9 **Relationship between an "crime rate" indicator and criteria**

Criteria	Objective
Table 6.5 Sustainable goals:	
Sustainable goal number 2	Intra-generation equity
Sustainable goal number 8	Individual well-being
Table 6.6 General selection criteria:	
General selection criterion (letter F)	Understandable by potential users
General selection criterion (letter H)	Comparable with other objectives
General selection criterion (letter I)	Cost effective
General selection criterion (letter J)	Without ambiguities
General selection criterion (letter K)	Attractive to the media
Table 6.7 Approach:	
Approach	Relation with "condition"
Table 6.8 Areas participation	
Social	Relation with "safety"

Solving the problem using Mathematical Programming

Goal: To identify eight sustainable indicators out of a total of 16.
Subject to the following attributes:

- *Eight* sustainability goals (1,2,3,4,5,6,7,and 8);
- *Seven* general selection criteria (B,F,G,H,I,J,K);
- Environment Canada framework (C,S,R);
- Areas participation;
- *Twenty five* minimum values (thresholds for attributes).

Additional restrictions:

It is established that there must be as a minimum:

- *One indicator* representing each area (in some cases this is not required);
- *Three indicators* complying with the sustainability objectives;
- *Five indicators* complying with general selection criteria.

The problem is then posed as detailed in Table 6.10. Applying MP to the above mentioned conditions, the system produces the results indicated in Tables 6.11 and 6.12.

Analysis of results

In order to see how the result satisfies attributes, it is necessary to compare columns *Required* and *Found* (Tables 6.11 and 6.12), which indicate how many indicators match a chosen criterion. In all cases the values shown in column *Found* match or exceed the values in column *Required.*

a. It is revealed that for the specified *sustainability goals, areas, and Environment Canada framework*, MP has chosen indicators: 2, 15,16.
b. For *general selection criteria,* MP has chosen indicators: 1,2,3,7,12. Observe that the total number of indicators found matches the number of indicators required (8), as well as their composition (3) for sustainability objectives, and (5) for general selection criteria.
c. All required areas are represented.
d. The three required components for the approach are represented.
e. Regarding the number of indicators for each attribute (sustainability objectives, general selection criteria, areas participation, and Environment Canada framework, check that out of 19 attributes, 19 (100 %), are satisfied.

Table 6.10 *Indicators and criteria*

CRITERIA	1	2	3	4	5	6	7	8	9	10	11	12	13	14	15	16			Indicators representing:
1	1	1	1		1					1		1	1			1	Choose at least	2	inter-generation
2				1		1			1		1			1		1	Choose at least	1	intra-generation
3	1	1	1							1		1	1			1	Choose at least	1	minimum impact
4	1	1	1													1	Choose at least	1	living off interest
5		1									1					1	Choose at least	1	non-renewable resources
6							1	1	1			1	1		1		Choose at least	1	long term economic development
7							1										Choose at least	0	diversty
8				1	1	1			1					1	1		Choose at least	1	individual well-being
B		1			1									1		1	Choose at least	1	representative
F	1	1	1	1	1	1	1	1	1	1	1			1	1	1	Choose at least	2	understandable by users
G	1																Choose at least	1	comparable with objectives
H	1	1	1	1	1	1	1	1	1	1	1	1	1	1	1	1	Choose at least	3	comparable with other objectives
I	1	1	1	1	1	1	1	1	1					1	1		Choose at least	3	cost-effective
J	1	1	1	1	1	1	1	1	1	1	1			1	1	1	Choose at least	3	with unabiguities
K	1	1	1	1	1	1		1	1	1	1			1	1	1	Choose at least	3	attractive to the media
Environment	1	1									1	1	1				Choose at least	1	environment
Social			1														Choose at least	0	economic development
Economics					1										1		Choose at least	1	public health
Env. - Social						1											Choose at least	0	safety
Env. - Economics							1	1	1					1			Choose at least	0	economy
Eco. - Social										1							Choose at least	0	people participation
Env. - Eco - Soc.				1													Choose at least	0	education
Condition	1					1	1		1				1			1	Choose at least	1	condition
Stress		1	1	1	1			1						1			Choose at least	1	stress
Response										1	1	1			1		Choose at least	1	response

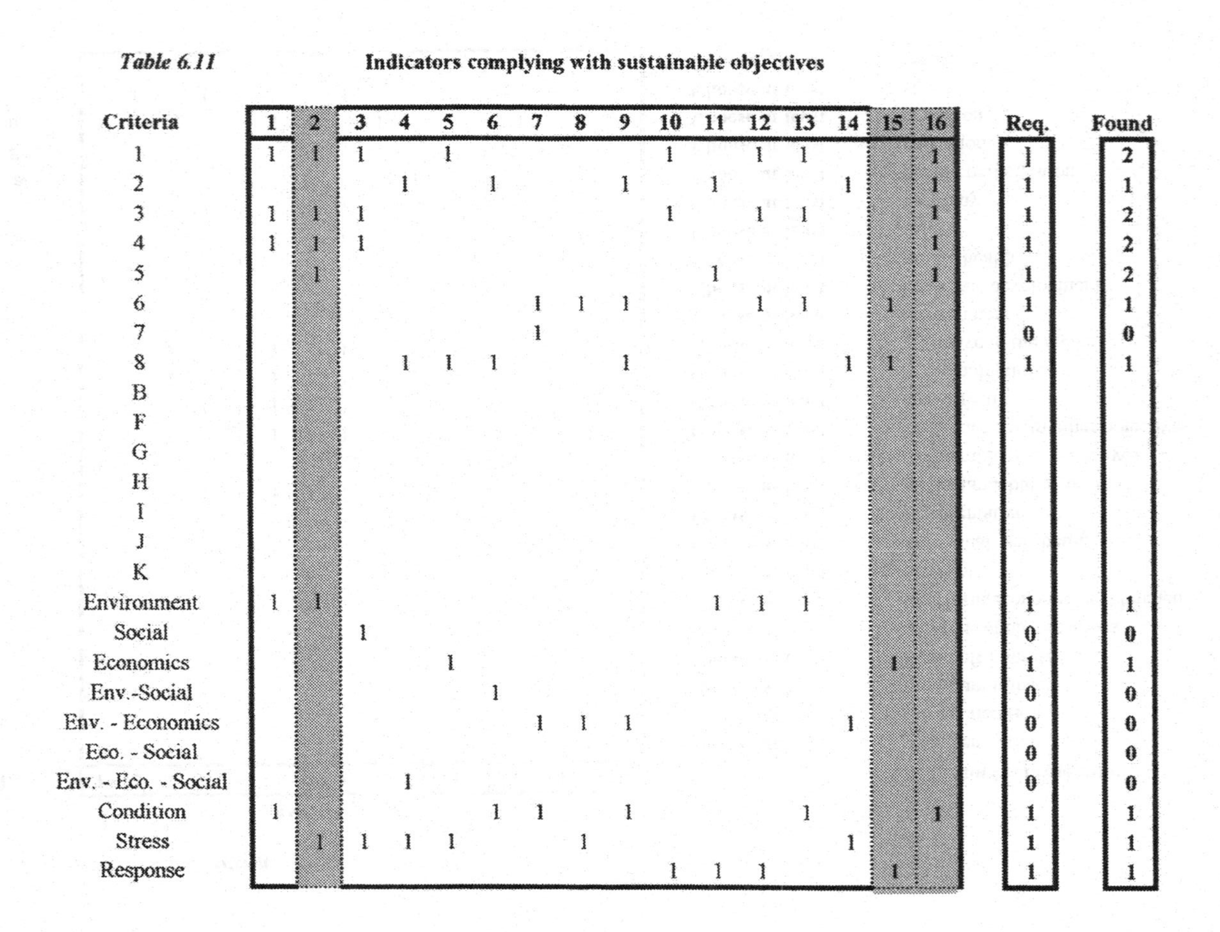

Table 6.11 **Indicators complying with sustainable objectives**

Criteria	1	2	3	4	5	6	7	8	9	10	11	12	13	14	15	16	Req.	Found
1	1	1	1		1					1		1	1			1	1	2
2				1		1			1		1			1		1	1	1
3	1	1	1							1		1	1			1	1	2
4	1	1	1													1	1	2
5		1									1					1	1	2
6							1	1	1			1	1		1		1	1
7							1										0	0
8				1	1	1			1					1	1		1	1
B																		
F																		
G																		
H																		
I																		
J																		
K																		
Environment	1	1									1	1	1				1	1
Social			1														0	0
Economics					1										1		1	1
Env.-Social						1											0	0
Env. - Economics							1	1	1					1			0	0
Eco. - Social																	0	0
Env. - Eco. - Social				1													0	0
Condition	1					1	1		1				1			1	1	1
Stress		1	1	1	1			1						1			1	1
Response										1	1	1			1		1	1

Table 6.12 **Indicators complying with general selection criteria**

	1	2	3	4	5	6	7	8	9	10	11	12	13	14	15	16	Req.	Found
1																		
2																		
3																		
4																		
5																		
6																		
7																		
8																		
B		1			1									1		1	1	1
F	1	1	1	1	1	1	1	1	1	1	1			1	1	1	2	4
G	1																1	2
H	1	1	1	1	1	1	1	1	1	1	1	1	1	1	1	1	3	5
I	1	1	1	1	1	1	1	1	1					1	1		3	4
J	1	1	1	1	1	1	1	1	1	1	1			1	1	1	3	4
K	1	1	1	1	1	1		1	1	1	1			1	1	1	3	3
Environment																		
Social																		
Economics																		
Env.-Social																		
Env. - Economics																		
Eco. - Social																		
Env. - Eco. - Social																		
Condition																		
Stress																		
Response																		

6.8.2 Case study --- Selection of urban indicators and projects

Background information

In an actual study for a city, *76 urban indicators* were pre-selected in the following areas:

- Environment
- Social
- Economics
- Socialeconomics-Environment
- Infrastructure
- Socialeconomics
- Citizens participation
- Municipal administration
- Municipal organization

The study called for a selection of 20 final indicators complying with the following conditions:

- Provide the maximum quantity of reliable information with the minimum number of indicators;
- To represent a wide scope of fields and activities;
- To be cost effective;
- To represent sustainable issues.

These conditions amounted to a total of 26 restrictions to be met.

A further study with *27 pre-selected projects* provided by the City Hall, together with 50 criteria, plus 29 indicators, comprised the city itself and eight satellite municipalities. It is important to mention that in this study some projects were linked each with another, as in projects involving more than one city, for instance sewage and road projects. Multicriteria analysis was employed and a *final selection of projects* was made. It is interesting to point out that *two of the 27 pre-selected projects were rejected by the model because they did not comply with sustainable requirements.* But the balance of projects selected met the terms imposed by all the restrictions posed, including budgetary constraints for each city, as well as all the threshold values imposed by sustainable indicators. In other words, the projects selected *were feasible from the economic, social, environmental and sustainability points of view.*

6.8.3 Case study --- Slum upgrading

Background information

The exercise consists in making a selection in an urban area using multicriteria analysis with the objective of providing the benefits of water and sewerage to the maximum number of people and at the lowest cost.

Assumptions:

1. The provision of basic infrastructure is analyzed, that is drinking water and sanitation as well as electricity, to communities that do not have these services.
2. Conditions vary from one urban area to another, for instance the cost of providing these services per hectare can show a large variation.
3. Priorities have also been established based on certain activities in some areas. As an example, there is an urban center which is an important hub to automotive and agriculture machinery repairs.
4. Financing is assumed from some lending institution. *Population willingness* and *capacity to pay* are especially considered in this example, as well as limits for investment per hectare for upgrading, and limits for per capita annual payments.
5. Five urban centers of different size are considered.

Objective:

Maximize the number of people taking advantage of infrastructure upgrading.

Decision variables:

Annual interest rate, in percentage:	4
Repayment period, in years:	5
Density, in persons/ha.:	470
Maximum total payment per capita:	51

Tables 6.13 to 6.17 condense information for the different areas. All tables are identical in the number of areas and in criteria, but of course data changes for each one. Thus, the cost of upgrading is different for each area *because the cost per hectare varies in accordance with the topography of the site.* Number of hectares to be developed depends not only on the size of each area but also on a *minimum area compatible with economies of scale*, that is, the minimum size of a profitable operation.

Both, *numbers of hectares and population define the density* for each site. There is a value of maximum investment per area that conforms to an equitable distribution of available funds. Each area has also particular values for their ability to pay, which depends mainly on the social and economic characteristics of each one. The same applies to the cost per capita and annual payment per hectare. The same interest rate and years for repayment applies to all areas, but the *corresponding cash flows will be different for each site.*

Table 6.13 **Area 1**

Years		**1**	**2**	**3**	**4**	**5**
Criteria						
Cost of upgrading ($/ha)	16,400					
Min. number of hectares to be developed	26					
Max. number of hectares to be developed	35					
Density (persons/ha)	427					
Total population in the area	14,945					
Max. total investment (US$)	574,000					
Ability to pay (US$/month-person)	0.72					
Annual interest rate	0.04					
Cost per capita ($)	43					
Annual payment per hectare ($/ha)	3,684					
Cash flow (investment, annual payments) per hectare	-16,400	3,684	3,684	3,684	3,684	3,684
NPV	0					

Table 6.14 **Area 2**

Years		**1**	**2**	**3**	**4**	**5**
Criteria						
Cost of upgrading ($/ha)	20,500					
Min. number of hectares to be developed	54					
Max. number of hectares to be developed	72					
Density (persons/ha)	450					

Total population in the area	32,400					
Max. total investment (US$)	1,476,000					
Ability to pay (US$/month-person)	0.85					
Annual interest rate	0.04					
Cost per capita ($)	51					
Annual payment per hectare ($/ha)	4,605					
Cash flow (investment, annual payments) per hectare	-20,500	4,605	4,605	4,605	4,605	4,605
NPV	0					

Table 6.15 **Area 3**

Years		**1**	**2**	**3**	**4**	**5**
Criteria						
Cost of upgrading ($/ha)	17,800					
Min. number of hectares to be developed	71					
Max. number of hectares to be developed	94					
Density (persons/ha)	396					
Total population in the area	37,224					
Max. total investment (US$)	1,673,200					
Ability to pay (US$/month-person)	0.84					
Annual interest rate	0.04					
Cost per capita ($)	50					
Annual payment per hectare ($/ha)	3,998					
Cash flow (investment, annual payments) per hectare	-17,800	3,998	3,998	3,998	3,998	3,998
NPV	0					

Table 6.16 **Area 4**

Years		**1**	**2**	**3**	**4**	**5**
Criteria						
Cost of upgrading ($/ha)	21,000					
Min. number of hectares to be developed	32					
Max. number of hectares to be developed	42					
Density (persons/ha)	517					
Total population in the area	21,714					
Max. total investment (US$)	882,000					
Ability to pay (US$/month-person)	0.75					
Annual interest rate	0.04					
Cost per capita ($)	46					
Annual payment per hectare ($/ha)	4,717					
Cash flow (investment, annual payments) per hectare	-21,000	4,717	4,717	4,717	4,717	
NPV	0					

Table 6.17 **Area 5**

Years		**1**	**2**	**3**	**4**	**5**
Criteria						
Cost of upgrading ($/ha)	22,500					
Min. number of hectares to be developed	53					
Max. number of hectares to be developed	70					
Density (persons/ha)	512					
Total population in the area	35,840					
Max. total investment (US$)	1,575,000					
Ability to pay (US$/month-person)	0.82					
Annual interest rate	0.04					
Cost per capita ($)	49					
Annual payment per	5,054					

hectare ($/ha)						
Cash flow (investment, annual payments) per hectare	-22,500	5,054	5,054	5,054	5,054	5,054
NPV	0					

With this information Table 6.18, the database, can be prepared. As usual areas are in columns while criteria are in rows. A number "1" is placed at the intersection of an area with a criterion. Thus, the intersection of "Area 1" and "Hectares" indicates that there is a relationship between them. To each row corresponds a threshold value which is related with the corresponding criterion.
For instance, in the first row the threshold value is 35, because this is the maximum size in hectares of Area 1, and as a consequence indicates the *maximum quantity* of hectares to be developed. This is indicated by the "<" sign.

The intersection of the second row and the first column also has a 1, as before, but observe that the threshold value has changed. It is now "26", and indicates that as a *minimum 26 hectares* are to be developed, and this is indicated by the ">" sign. The last row shows a series of "1s" and a threshold value of $ 6,180,200 with the "<" sign indicating that said amount is the *maximum available funding*.

Table 6.18 **Database**

Urban areas	**1**	**2**	**3**	**4**	**5**			
Density	427	450	396	517	512			
Ability to pay (annuity)	3,684	4,605	3,998	4,717	5,054			
Weight in accordance with infra-structure efficiency	5,000	5,000	9,000	6,000	7,000			
Units						**Results from computation**	**Sign**	**Thresholds**
Hectares	1					35	<	35
Hectares	1					35	>	26
Hectares		1				72	<	72
Hectares		1				72	>	54
Hectares			1			94	<	94
Hectares			1			94	>	71

Hectares				1		38	<	42
Hectares				1		38	>	32
Hectares					1	64	<	70
Hectares					1	64	>	53
US$	1					574,000	<	574,000
US$		1				1,471,238	<	1,476,000
US$			1			1,673,200	<	1,673,200
US$				1		801,818	<	882,000
US$					1	1,445,801	<	1,575,000
US$/ha	1					43	<	51
US$/ha		1				51	<	51
US$/ha			1			50	<	51
US$/ha				1		41	<	51
US$/ha					1	45	<	51
People/ha	427					14,945	<	16,450
People/ha		450				32,295	<	33,840
People/ha			396			37,224	<	44,180
People/ha				517		19,740	<	19,740
People/ha					512	32,900	<	32,900
Financing (US$)	1	1	1	1	1	**5,966,057**	<	**6,180,200**

The results are depicted in Table 6.19. Notice that the column *Results from computation* in Table 6.18 indicates a total expenditure of $ 5,966,057 which is *less* (as it should be), than the maximum amount of available funding of $ 6,180,200.

Table 6.19 **Results**

Urban areas	**1**	**2**	**3**	**4**	**5**
Density	427	450	396	517	512
Ability to pay (annuity)	3,684	4,605	3,998	4,717	5,054
Weight in accordance with infrastructure efficiency	5,000	5,000	9,000	6,000	7,000
Hectares to be developed	**35**	**72**	**94**	**38**	**64**
Total number of people	**14,945**	**32,295**	**37,224**	**19,740**	**32,900**
Total cost	**574,000**	**1,471,238**	**1,673,200**	**801,818**	**1,445,801**

Comments on results:

In Table 6.19, the row *Hectares to be developed* shows the optimal distribution of land, and the row *Total number of people* indicates the maximum quantity of people to receive the benefits. The *Total cost* row indicates the amounts allocated to each area. Comparing the amount of funds to be used with the amount available, it results in a very high rate of use of the available fund (97 %), which is remarkable considering all the restrictions imposed. Also, out of a total of 142,123 people, the system allocated 137,104, and out of a maximum amount of 313 hectares, 303 will be urbanized..

Note: To compare values as well as for monitoring purposes some "liveability" indexes can be used, such as:

- Percentage of people with access to energy.
- Percentage of people with access to clean water (communal faucets).
- Percentage of people with access to sanitation (latrines).
- Percentage of people with secure tenure.

The numerical value for each of these indexes could have been incorporated in these examples as additional thresholds.

6.9 Monitoring

The United Nations in its UNEP EIA Training Resource Manual, (see Bibliography under UNEP), define Monitoring as: *The systematic collection of environmental data through a series of repetitive measurements.* To monitor in the EIA scenario means the following activities:

- Determine what is to be monitored;
- Choose frequency of checking;
- Decide on the size of the sample;
- Check that effects are within the established limits and thresholds;
- Determine trends from impacts;
- Identify additional consequences;
- Appraise how well mitigation measures are working;
- Detect unforeseen effects.

A well done EIA study most likely will *determine the effects* that a certain project will create in a *spatial* and *temporal basis*. It will also provide an estimate of the *magnitude and concentration* of these effects, and hopefully will also give a good idea of how these impacts will affect human health and the ecosystem. However, if a mechanism is not established to check these expected changes, and most important to *control that they remain within certain limits or thresholds*, then the whole purpose of the study is lost.

Besides, sometimes it is almost impossible to determine all the direct, indirect and cumulative effects that a project or a set of projects will trigger. There are many things that we don't know about and many unexpected reactions that will surprise us. Again, we need to have a control of these unforeseen circumstances once they are evident. For all of the above it is necessary to exert a *control or a monitoring* after a project is operational. This is also routine in many other aspects other than EIA, as we monitor the state of our health, the mechanical status of our car, and we look for deterioration signs in our house.

Internet references for Chapter 6

Indicators
Title: *Urban indicators program*
City Managers' Association Gujarat
Conceptual information.
http://www.cmag-india.org/programs_urban_indi_prog.htm

Title: *Appendix A: Urban Data Tables*
World Resources 1996--1997
Urban indicators and information of cities around the world.
http://www.wri.org/wri/wr-96-97/wr96dtur.pdf

Title: *Global Urban Observatory*
United Nations Human Settlements Programme
Information about the Global Urban Observatory (GUO)
http://www.unchs.org/programmes/guo/

Title: *Most frequently used urban indicators*
Authors: Valentinelli Alessandra & Anna Crimella
This is a research work where the authors have determined which are the most frequently used urban indicators and rank them. It is a very useful work.
http://esl.jrc.it/dc/urban_indicators.htm

Strategic Planning
Title: *Urban development study for the extended zone of Guadalajara, according to indicators of sustainability, Mexico, 2002*
Authors: Aguilar O, Arredondo J.C., Munier N., Calvo N., D'Urquiza A.
http://challenge.vegasys.net/search_view.asp?Id=222

Title: *A tool for urban strategic planning*
Author: Munier Nolberto
The use of a methodology called SIMUS for defining an urban strategy http://wbln0018.worldbank.org/UrbanCalendar/urban.nsf/0/8be4b987fd22782485256a220060f54d?OpenDocument

Title: *Summary of urban competitive assessment*
The World Bank Group
http://www1.worldbank.org/nars/ucmp/UCMP/step_two_urban_competitiveness.html

Title: *Urban development*
The World Bank
http://www.worldbank.org/urban/

Title: *Urban waste management --- Strategic planning guide for municipal solid waste management*
Authors:
David Wilson, Andrew Whiteman and Angela Tormin, Environmental Resources Management for the Collaborative Working Group, 2001.
http://wbln0018.worldbank.org/External/Urban/UrbanDev.nsf/0/349F2CDAE6E96C6285256B3A007DB3D1?OpenDocument

APPENDIX : MATHEMATICAL BACKGROUND TO UNDERSTAND THE FUNDAMENTALS OF MODERN TECHNIQUES FOR PROJECT APPRAISAL

A.1 Introduction

Some of the most important techniques used in EIA are based in mathematical notions. If you are familiar with the following concepts you can skip this section. The topics very briefly commented here, *without mathematical rigor* are intended for people with no mathematical background, and as a consequence expressed in plain language.

Subjects covered:

- Regression analysis;
- Correlation analysis;
- Matrices;
- Eigenvalues and eigenvectors;
- Link between matrices and correlation;
- Linear transformation of the unit circle;
- Relationship between the ellipse and the correlation matrix;
- Mathematical foundation of the Analytic Hierarchy Process;
- Principal components analysis;
- Factor analysis;
- Mathematical foundation for Mathematical Programming;
- Mathematical foundation for Input/Output analysis.

A.1.1 Regression analysis

Regression analysis provides a systematic technique for estimating with confidence limits, the *unspecified constants* from a new set of data, or for testing if the new data is consistent with the hypothesis. The general method used in estimating a population regression curve from sample data is the method of *least squares.* The sample regression curve of **y** on **x** of a given degree is the curve among those of that degree that minimizes the sum of squares of vertical **y** deviations of the observed points from the curve.

If the relationship betwen the variables is linear, the result is a *linear regression equation* which shows the average **y** values (which is the *dependent* variable) for fixed **x** values (which is the *independent* variable), and as a straight line (the *regression line*). The equation has a constant term and a coefficient for the **x** value. This coefficient is the *regression coefficient* and the *slope* of the *regression line.*

Regression analysis finds the relationship between dependent and independent variables, and determines if this relationship is linear, parabolic, hyperbolic, or of some other type. The strength or the weakness of this relationship is measured by the *correlation coefficient* (see section A.1.2).

If we represent in a graph with x and y coordinates the location of each x-y pair we get a point. The representation of all points forms a *cloud of points* or a *scatter plot.* See Figure A.1.

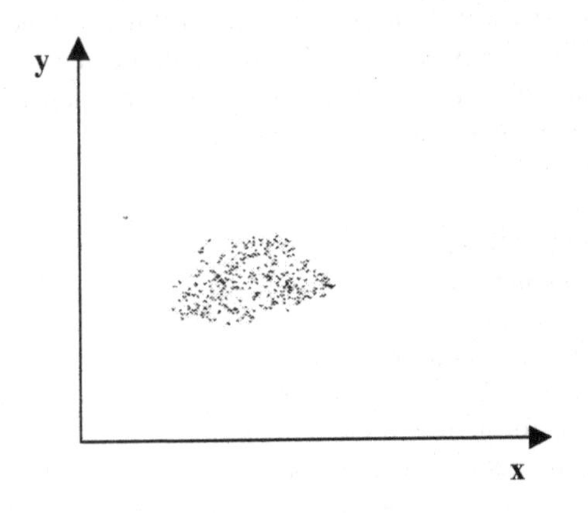

Figure A.1. **Scatter plot shows no correlation between variables**

Is it possible for us to guess if x and y are in some way related?

Well, in this case it is impossible since we cannot see a pattern identifying some kind of relationship, that is, nothing suggests the existence of an association, materialized by means of a certain type of line between both variables, that could indicate how y *varies* for *different values of x.*

A.1.1.1 Example --- Analysis of an urban transportation corridor

Suppose now that we are analyzing a transportation corridor in a city. There are statistics between 1995 and 2002 about the *ridership* (that is the number of passengers who ride an urban transportation system in that corridor), and the *quantity of buses* serving it, and we want to determine the existent kind of relationship, if any, between these two variables. Ridership is expressed as $\mathbf{x_1}$ and is the *independent* variable, while number of buses is the *dependent* variable and indicated as **y**.

Table A.1 shows these statistics and Figure A.2 is the graphical representation of this relationship.

Table A.1 **Data on urban travel in a corridor**

Year	Quantity of buses	Ridership (x 000)
	y	x_1
1995	563	58,976
1996	542	65,375
1997	547	68,865
1998	**642**	**76,448**
1999	732	83,504
2000	835	92,007
2001	852	94,062
2002	868	87,291

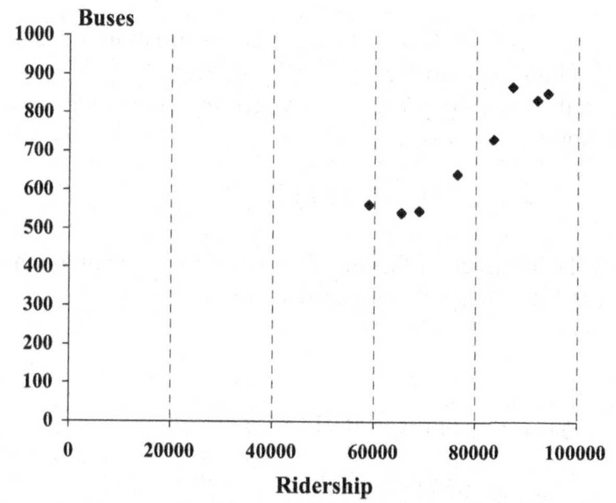

Figure A.2. **Relationship between ridership and number of buses**

Examination of this graph indicates some kind of relationship between $\mathbf{x_1}$ and **y**, since we can observe that when ridership increases, also increases the number of buses, and in fact we could draw by hand a line across the points showing this relationship, but it would be convenient to know the *analytical form or equation of this line,* for knowing one of the variables we can determine the other. This can be done through *regression analysis,* which minimizes the squares of the distance between each point and the corresponding point on the line, i.e., the two points which are on the same vertical line.

Any book on statistics can explain how to calculate this equation, but it is easier to use some of the add-ins included in spreadsheet programs which can ver easily compute and draw it. The equation has the following form:

$$y = a + b\, x_1$$

It is a linear equation and **a** and **b** are the values of the regression equation that can be found employing one of the above mentioned add-ins. What is the meaning of these two values?

"a" is a constant and shows the value where the regression line intersects the vertical axis (y).

"b" is a coefficient that indicates the inclination or slope of the regression line with respect to the horizontal axis (x_1).

Using the data of Table A.1 we can get the numerical values of the regression equation.

$$y = -\ 111.6 + 0.013\, x_1$$

Just to *test* the accuracy of this equation say that we want to determine the number of buses for year 1998, with a ridership of 76,448.

Then:

$$y_{1998} = -\ 111.6 + 0.013 \times 76{,}448$$

$$y_{1998} = 676 \text{ buses}$$

As per Table A.1 the actual value for that year was 642 buses so there is a discrepancy of 676 - 642 = 34 buses, which is explained because the two variables are not perfectly correlated (see section A.1.2). Now, does the fact that we can determine **y** from $\mathbf{x_1}$ mean that there is a cause-effect relationship between both? In other words does it mean that **y** happens because of $\mathbf{x_1}$?

Not necessarily, both variables can go up or down simultaneously and this does not mean that there is a cause-effect relationship between them. Just think that we can have two variables such as the number of people going to the movies and the increase in bird's population in a nearby park. Both can be increasing even at the same rate, and obviously one fact has nothing to do with the other, and albeit both happen to grow with time, they are not related.

Notice that entering in the regression equation $\mathbf{x_1}$ (independent value), it is possible to compute value **y** (dependent value), and the two variables become in *reality one*. This is very important as we will see when commenting about Factor Analysis. (see section 3.2).

Multivariate linear regression

What happens when there are more than two variables, that is, what if for instance a new variable x_2 is added? The concept and the procedure is the same, but instead of representing the problem in two dimensions (y and x_1) it is necessary to work now with three dimensions (y, x_1 and x_2). We still will get scatter points but in this case it will be in three dimensions, so in lieu of a two-dimensional cloud as it was before, now there will be an *elongated ball* similar to a football, and because we are working in a three dimensional space there is no longer a regression line but a *regression plane.* Each new variable adds a new dimension to the problem so, with say seven variables there will be seven dimensions. Is that possible?

Yes, it is in a mathematical sense, albeit obviously nobody can visualize more than the familiar three dimensional space.
Following with the transportation example, we can add now two new variables: *distance travelled* (x_2) and *operating costs* (x_3) which data is depicted in Table A.2, so now the problem lies in a four- dimensional space.

Again, **y** is the dependent variable and $\mathbf{x_1}$, $\mathbf{x_2}$ and $\mathbf{x_3}$ the independent variables, and it is at this time possible to calculate for the same year the regression equation considering also variables x_2 and x_3. This is called *multivariate regression,* and any statistics software will render the following equation for the estimate of the number of buses for year 1998.

Table A.2 **Complete data on urban travel in a corridor**

Year	Quantity of buses	Ridership (x 000)	Distance traveled (Km) (x 000)	Operating costs (x 000)
	y	x_1	x_2	x_3
1995	563	58,976	27,648	17,554
1996	542	65,375	29,367	19,219
1997	547	68,865	30,000	20,800
1998	**642**	**76,448**	**34,747**	**26,722**
1999	732	83,504	40,462	36,914
2000	835	92,007	50,168	50,384
2001	852	94,062	54,424	62,188
2002	868	87,291	54,553	71,781

The multivariate equation will be now:

$$y_{1998} = 150.8 - 0.000048\ x_1 + 0.01527\ x_2 - 0.00164\ x_3$$

Replacing values for the independent variables, from Table A.2:

$$y_{1998} = 150.8 - 0.000048 \times 76{,}448 + 0.01527 \times 34{,}747 - 0.00164 \times 26{,}722$$

$$y_{1998} = 634 \text{ buses}$$

Notice that the difference from the true value is 642 - 634 = 8

The *margin of error was greatly reduced* because the inclusion of the two added variables *increased the accuracy* of the prediction. The coefficients affecting independent variables x_1, x_2 and x_3, are called *partial regression coefficients,* and an inspection of these values can tell us which of the independent variables is more significant as a determinant of **y**. However, this comparison cannot be made directly because the units of measure are different, for that reason the regression coefficients have to be converted to *normalized coefficients.* This is a simple procedure but it is beyond the scope of this book, and can be found in any statistics book.

Forecasts

The same computation used above to test the accuracy of the procedure can be utilized for forecasts, which of course is the whole purpose. For instance, employing different techniques the ridership for year 2006 can be estimated, together with estimates for the other two variables and, using the above equation, it is possible to compute the number of buses that will be needed in 2006. Suppose for instance that the estimate for the independent variables for year 2006 is, in thousands:

Ridership:	90,438
Distance traveled:	55,896
Operating costs:	77,421

The estimated number of buses will then be:

$$y_{2006} = 150.8 - 0.000048 \times 90{,}438 + 0.01527 \times 55{,}896 - 0.00164 \times 77{,}421$$

$$y_{2006} = 873 \text{ buses}$$

A system of two variables can be easily solved by hand, and with a little more effort if it has three variables, but with more variables, even if the procedure is the same it becomes too complicated and the probability of making mistakes increases exponentially. For that reason add-ins from the spreadsheet programs or statistics packages must be used.

A.1.2 Correlation analysis

Regression has been defined and therefore one variable can be computed as a function of another. Now, what is the degree of association between these two variables? In other words, having estimated **y** as a function of $\mathbf{x_1}$ how *exact an approximation* is this estimate?

A first idea can be obtained just with a look at the cloud of points -- but of course it can only be done in a two--- or in a three-dimensional system --- and then it is possible to observe *how far away* are the points of the cloud from a regression line drawn by hand or using the regression equation. If the cloud of points is as depicted in Figure A.1, it is obvious that a regression line can have any direction, and as a consequence it is not possible to make even a guess of where it will lie, so it appears that the relationship between both variables is very poor, or inexistent.

Assume now for a certain process involving two variables the corresponding scatter points as depicted in Figure A.3. Here it can be seen that the points group around and are near the regression line, meaning that there is probably a close relationship between both variables. But this is a *very subjective assessment*, so one wonders if there is some way that could be used to *put a value on this closeness or lack of it.* Yes, there is a measure, it is called *the correlation coefficient,* varies from 0 to 1, and is identified as **r**.

Figure A.3. **Cloud very close to the regression line**

When it is zero or close to zero, maybe 0.10 or 0.12, then the relationship between both variables is *poor* or *inexistent*. When it has a high value such as 0.94 for instance, the relationship is strong and the *width* of the cloud will be very *thin* and approaching to the regression line. What happens if the cloud is so thin that it blends into the regression line and there is no difference between it and the cloud?

Then the correlation is perfect and equal to 1. As an example, assume that there is a river where many food processing plants discharge their liquid wastes without any treatment. The degree of contamination is measured by the content of BOD_5, which is the Biological Oxygen Demand (see Glossary). When the BOD_5 content increases, the fish population in the river --- at least in the area close to where the discharges take place --- decreases because they do not have enough oxygen dissolved in the water to breath.

When the level of BOD_5 is so high that life is impossible, then there are no fishes and we can have a perfect correlation between *impact* (the discharge of contaminant) and *response* (the fish population or the death of fishes). Again this *is not a cause-effect relationship*, because the death of the fishes could also *have other causes*.

The calculation of the correlation coefficient can be made by hand, but again, statistical packages or the use of spreadsheet programs can calculate it very easily. In the transportation example used to illustrate regression analysis (A.1.1.) we first analyzed the relationship between $\mathbf{x_1}$ and $\mathbf{y}$. The correlation coefficient in this case is $\mathbf{r} = \mathbf{0.94}$, which, as expected, suggests a *strong relationship between ridership and number of buses*.

To calculate now the simultaneous relationship between $\mathbf{y}$ and $\mathbf{x_1}$, $\mathbf{x_2}$ and $\mathbf{x_3}$, it is necessary to use what is called the *multiple coefficient of determination* ($\mathbf{R^2}$). In this example, since it involves one dependent and three independent variables, we use the notation $\mathbf{R^2_{y.123}}$, and with a value of **0.9795**.

Again, as in the regression case, it can be seen how the introduction of the other two variables *explains almost in a 98 per cent the variation* of $\mathbf{y}$ due to $\mathbf{x_1}$, $\mathbf{x_2}$ and $\mathbf{x_3}$.

A.1.3 Matrices

A matrix is simply a table with columns and rows, with this format:

$$\begin{bmatrix} 6 & 9 \\ 8 & 5 \end{bmatrix}$$

Matrices are studied in linear algebra and are extremely important in practically all aspects of science. They are used everyday to solve problems such as the computation of the stress suffered by a body subject to a load, in sociological interactions, and for complex computations in physics such as those pertaining to quantum mechanics. Matrices are not new but have increasing utility thanks to the use of computers which perform --- if not complicated --- at least long and cumbersome calculations, for which they are ideally suited.

There are literally hundreds of books on matrices and their different classes and applications, from the very simple example used in solving High School problems to the very complicated associated with many different fields. In High School algebra, we learned how to solve a system of linear equations, say with three or four variables, solving first for one variable, then replacing expressions, solve for another and so on. This is called the Gauss-Jordan method, and it is still very much used.

Using matrices instead of the Gauss-Jordan method, the same result can be achieved but in an easier way. In EIA matrices are utilized in two very important tools such as the Analytical Hierarchy Process (AHP) (see section 5.8) and Mathematical Programming (MP) (see section 5.9). Matrices are also employed to explain capital concepts such as Factor Analysis (see section 3.2), an also useful tool for EIA. To solve matrices, mathematical software such as MathCAD® and others are utilized, and also some special add-ins incorporated in Excel®, Quatro Pro® and Lotus 123® spreadsheets.

A special and very valuable application of matrices for EIA is in Mathematical Programming, and for this, the three above mentioned software have dedicated add-ins.

A.1.4 Linking matrices and correlation

A *correlation matrix* is a table that shows the *correlation coefficients between variables*, as illustrated in Figure A.4.

$$\begin{bmatrix} 1 & 0.82 \\ 0.82 & 1 \end{bmatrix}$$

Figure A.4. **Correlation matrix**

Assume two variables x_1 and x_2, that can be represented in a Cartesian coordinate system such as shown in Figure A.5. On the horizontal axis x_1 are *all the values that the variable x_1 can take*, and in the vertical axis *all the values that the variable x_2 can assume.* Both axes are at 90 degrees, or perpendicular each to the other, and this also means that both variables are *independent,* so their correlation is *zero.* If you remember what it was said before regarding perfect correlation, then the x_1 axis represents the perfect correlation ($r^2 = 1$) *between the x_1 variable with itself,* and the *same for x_2.*

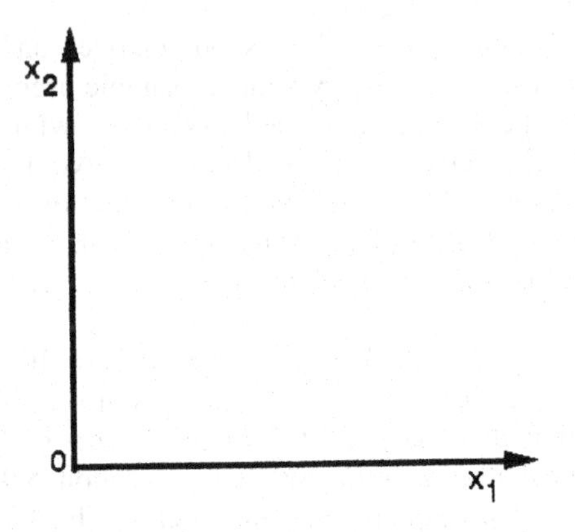

Figure A.5. **A coordinate system for two variables**

In Figure 6 we draw a circle with radius 1 and center at the origin of the coordinate system, then this *unit circle* represents the *loci of all points with radius 1.*

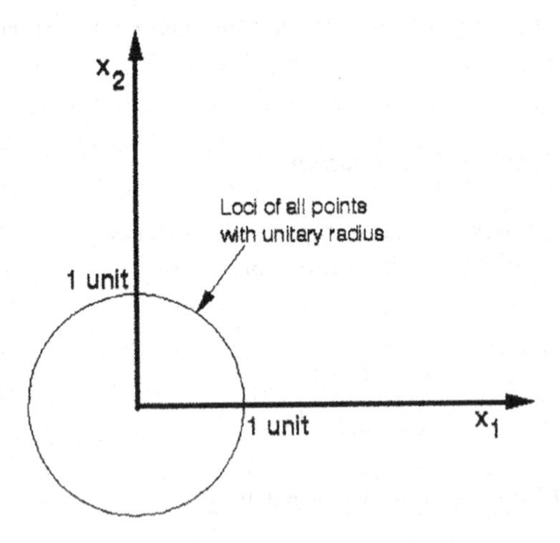

Figure A.6. **Loci of all points with unitary radius**

It is clear in a drawing, but how can this condition be expressed in a matrix? In this way:

$$\begin{bmatrix} 1 & 0 \\ 0 & 1 \end{bmatrix}$$

What does this mean?

Exactly what was said before.

The first column $\begin{bmatrix} 1 \\ 0 \end{bmatrix}$ i.e., x_1 column, indicates that the horizontal axis has a correlation coefficient of 1, and *zero when related with x_2.* The same for the second column that belongs to x_2 expressing that it has a perfect correlation of 1 with itself and *zero with x_1.* Because of the symmetry of its values, this matrix is called a *symmetrical matrix.*

A.1.5 Linear transformation of a unit circle

This is a mathematical operation which finds the "image" of a point when a *symmetrical matrix* is applied to it. This matrix finds the "transform" or "image" of each point.

What happens if a matrix is applied to this unit circle?

Then, to *each point of the circle will correspond now a new position or image,* within a Cartesian system, and the circle will transform in an *ellipse.* Because there are only two variables, this ellipse will have two axes, and these axes most probably will not coincide with the original x_1 and x_2 axes, so the ellipse is now the circle transformed by the linear transformation. See Figure A.7, with its axes having a different orientation that those of the circle.

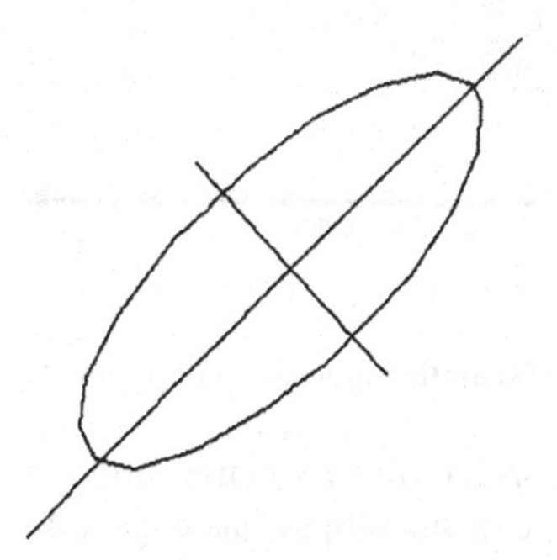

***Figure A.7.* Circle transformed into an ellipse**

A.1.6 Relationship between the ellipse and the correlation matrix

Suppose a correlation matrix with these values:

$$\begin{bmatrix} 1 & 0.7 \\ 0.7 & 1 \end{bmatrix}$$

It is a symmetrical matrix. Again the "1s'" are the correlations of each axis with themselves, while "0.7" is the correlation coefficient between them. We are now going to represent this correlation matrix in our coordinate system. To do that consider that the *first row* of the matrix establishes that there is 1 unit in the x_1 axis and 0.7 units in the x_2 axis. The intersection of both values determine a point A. Point B is determined in a similar manner considering x_2.

See Figure A.8., and observe that these two points A and B are on the periphery of the corresponding ellipse. A correlation factor of 0.7 is actually not too high, and shows a medium correlation between x_1 and x_2, and this determines the shape of the ellipse.

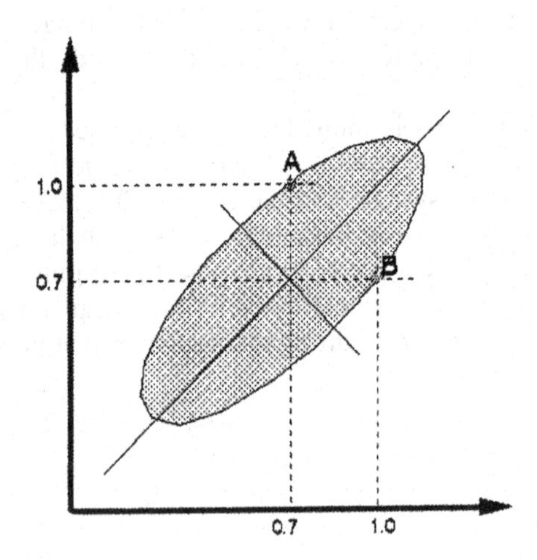

Figure A.8. **Representation of a correlation matrix**

Assume now that the correlation between x_1 and x_2 is 0.9, see Figure A.9. It can be seen that the shape of the ellipse has changed when compared with a correlation factor of 0.7, and it can be deduced that *the higher the correlation the thinner the ellipse.* Observe how the two points A and B are now closer to each other than before, and look at the shape of the ellipse which is now much narrower. What would happen in the case of a perfect correlation between x_1 and x_2?

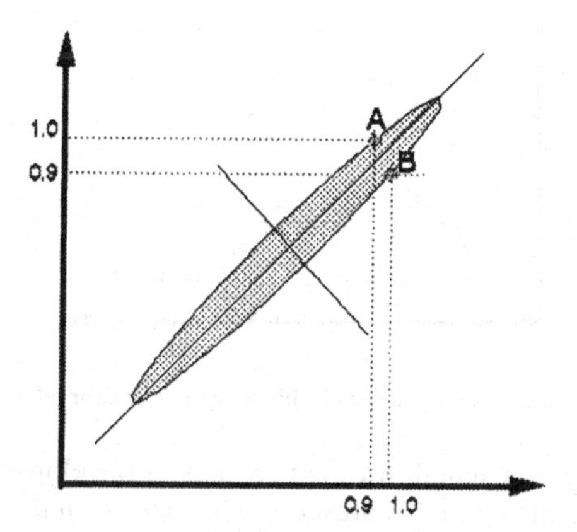

Figure A.9. **Change in the shape of the ellipse for increasing correlation**

The correlation matrix will then be:

$$\begin{bmatrix} 1 & 1 \\ 1 & 1 \end{bmatrix}$$

From the geometrical point of view it means that in this case *points A and B coincide*, and the *ellipse collapses into a line.* See Figure A.10.

In each case the resulting ellipse has two axes, a longitudinal axis and a shorter perpendicular axis. When axes in Figure A.9 are compared with the original axes in the unit circle (Figure A.6), it shows now that they are different in *length* and in *orientation, since the unit circle has converted into an ellipse.* The longitudinal axis in the ellipse exceeds the axis in the unit circle by a certain amount, and the ratio between both lengths is a *scale factor.* It is greater than 1 in the longitudinal axis and less than 1 in the perpendicular axis.

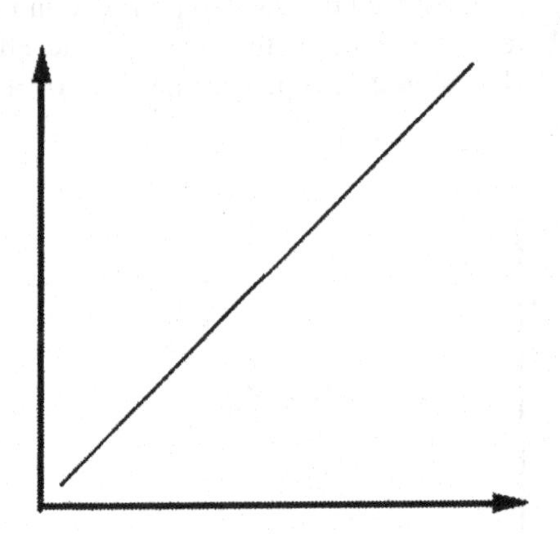

Figure A.10. **Collapse of the ellipse for perfect correlation**

The respective whole length for the new axis of the ellipse compared with the axis in the unit circle, represents *how much the unit circle has transformed.* These elongated values of lengths are called *eigenvalues.* The new axis of the ellipse are called *eigenvectors.* In a three dimensional scenario the figure will not become an ellipse but an *ellipsoid.* Figure A.11.

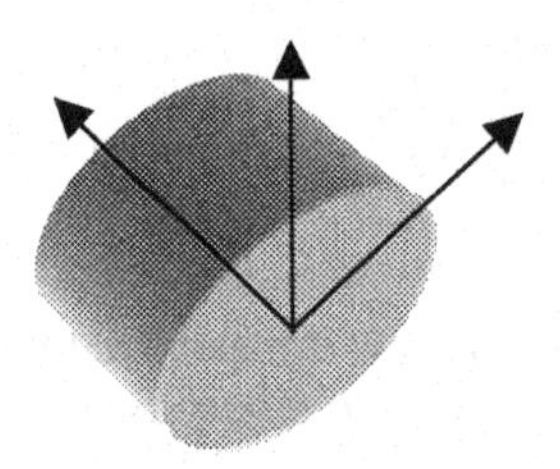

Figure A.11. **Ellipsoid**

As an everyday example take a familiar object in our also familiar three-dimensional world, for instance, a little piece of a rubber hose, perfectly cylindrical (this is our *unit circle*). Exert now some pressure on it (such as squeezing the hose with two fingers) (this is our *matrix*); what is the result?

The hose modifies its shape, taking an *ellipsoidal appearance,* as a result of the transformation due to the action of the force applied to it.

The length of the new axes of the hose will represent the scale factor, or *eigenvalues,* by which the original radius of the unit circle has been affected, and these two new axes will be in directions established by the *eigenvectors.*

As an example, in the case of a beam subject to a load, the eigenvalues would give us the stress produced by the load in the three directions of the beam (length, width and height). With these values we can know if the beam will stand the load. Eigenvalues and eigenvectors are a very important part of linear algebra and have countless applications in practically every field of science.

Where are they used in EIA?

They find their application in the Analytical Hierarchy Process (AHP) for the determination of the weights assigned to criteria and alternatives. They are also used in *Factor Analysis* (section 3.2 and Appendix, section A.5) to determine the underlying factors affecting a set of variables.

A.2 Mathematical foundation of the Analytical Hierarchy Process

This process, developed by the American mathematician Thomas Saaty, considers *pair-wise comparisons* between elements in order to find out their relative weight.

A.2.1 Example --- Analysis of a project with three criteria

Suppose that in a project there are three criteria to be used, as follows:

Criterion A : Air pollution
Criterion B : Financing
Criterion C : Public acceptance.

These are the criteria that will be used to select different alternatives of a project. Using a 1-9 scale, we compare pair-wise each criterion with each of the others. If criterion A is more important than criterion C, for example, and receives a 9 in the scale, then C related to A receives a 1/9. So, the following matrix, Table A.3, can be constructed:

Table A.3 **Criteria matrix**

Criteria	**A**	**B**	**C**
A	1	1/7	**9**
B	7	1	1/8
C	**1/9**	8	1

This is a *square asymmetrical matrix* and when applied to axes x_1, x_2 and x_3 an ellipsoid is obtained where the axes correspond to its length, width and height. How do we get the eigenvalues?

It is not worth developing here all the calculations to find the eigenvalues (see section A.1.6), since there is adequate software for it. In this case there will be three of them, and also three eigenvectors, one for each row. The eigenvectors give the directions of the axes corresponding to the eigenvalues. To find these values it is necessary to enter into calculations that are not difficult but that nowadays are solved very easily with computer programs. In this case, using these programs we will find the three eigenvalues which are denoted by the Greek letter lambda (λ), so we will get λ_1, λ_2, and λ_3. One of them will correspond to the largest axis of the ellipsoid, and the other two to the height and to the width axis respectively.

AHP chooses the largest lambda and for that value computes the eigenvectors. There is also a mathematical formula to find eigenvectors but we will use a simple formulation that was also utilized in our example in section 5.8.1. It consists in finding for each row the geometric media: The formula is simple, just multiply all the values of the row, and then extract the third root (in this case) of that quantity. Once this procedure is completed for every row, add them up to get the total. Next normalize the individual value by dividing each one by this total. In our example we have, Table A.4:

Table A.4 **Determination of criteria weights**

	Product of values in a row from Table A.3	**Result of product of values from column (1)**	**3rd root from column (2)**	**Values from column (3) divided by total**
	Column (1)	Column (2)	Column (3)	Column (4)
Row 1 - criterion A	1 x 1/7 x 9	1.2857	1.0873	**0.3618**
Row 2 - criterion B	7 x 1 x 1/8	0.875	0.9564	**0.3182**
Row 3 - criterion C	1/9 x 8 x 1	0.8888	0.9615	**0.3199**
		Total:	3.0052	

According to the values from column (4) these are the *weights for each criteria.* These weights are very important since they will affect the values determined for each project and each criteria. The same procedure can be followed to determine the relative weights corresponding to different alternatives or projects. As can be seen the AHP is based in sound mathematical principles, but naturally this was only a very brief introduction to the method. There are other aspects that have to be considered, but the essentiality of the process is as described. For more information please refer to the Bibliography, especially the work from Dr. Thomas Saaty. (see Bibliography- Saaty T.).

A.3 Foundation of Mathematical Programming

A methodology which is increasingly being used for Environmental Impact Assessment (EIA), is *Multicriteria Analysis.* As its name implies, this technique analyses projects, alternatives or options that are subject to many different criteria that pertain to different fields, such as the environment society, economics, diverse risks, etc. These criteria are then used to make the selection of projects considering how well each alternative matches them. There are several techniques utilizing this approach. Here the fundamentals of Linear Programming --- a part of Mathematical Programming --- are briefly described. This technique is nowadays widely employed on a daily basis in hundreds of applications, and its use will be explained by an example.

A.3.1 Example --- Analysis of a project that calls for the construction of an oil pipeline, with four alternatives or options.

Alternatives

The different alternatives arise because four different potential routes have been studied, each one linked with a different aspect, such as soil condition, citizens' preferences, environmental risk, crossing of protected natural areas, needs for tunneling, terrain slopes, cost, etc. The construction cost of each alternative has been calculated, as well as its operating and remediation costs.

The objective of this exercise is to select the best possible alternative in order to minimize the cost to the environment, and compatible with the monetary cost of the undertaking. A table or matrix is used to depict the alternatives in columns. These alternatives constitute a vector which we identify as "**X**".

Criteria

We can have a large quantity of criteria, identified as 1, 2, 3, ...,n. See Figure A.12. Each criterion is shown in a row and for each one threshold values can be established *(for "threshold standards" see section 3.1.12).*

Therefore, each alternative will be evaluated by **n** different criteria, with equal or dissimilar units of measure. At the intersection of an alternative (in a column) with a certain criterion (on a row), we place a number which indicates how well the corresponding alternative conforms to the respective criterion requirement. This value is called a *coefficient.*

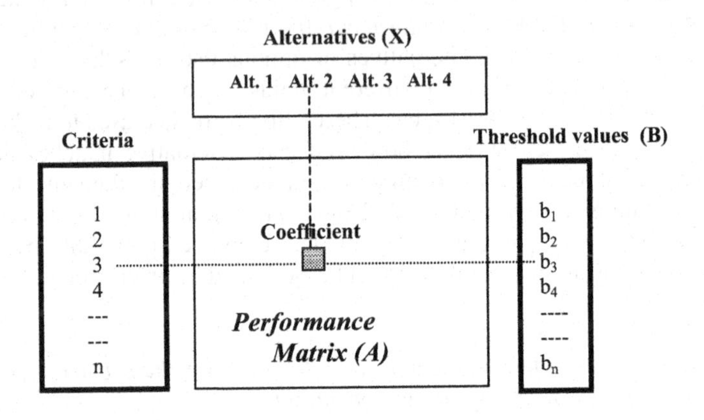

Figure A.12. **Performance matrix**

Performance matrix

The set of alternatives, criteria, coefficients, and threshold values constitutes what is called a *Performance Matrix,* since as it name implies, it measures the performance of all alternatives regarding criteria.

Analysis

If in the Performance Matrix, for criterion 1 and the four alternatives, these coefficients are specified --- which in general are indicated as a_{ij} --- where i = rows and j = columns, then it can be established that:

For the first criterion:

$$a_{11}\,x_1 + a_{12}\,x_2 + a_{13}\,x_3 + a_{14}\,x_4 = b_1$$

In other words: Adding up the products of each coefficient (a_{ij}) by each alternative (x_j), the result should be compared with the threshold value for that criterion (b_i).

According to the characteristics of the problem, any of the symbols, i.e., the *equal* symbol (=), the *greater than* symbol (>) and the *less than* symbol (<) can be assigned to each criterion. Sometimes a criterion uses the two symbols "<" and ">" especially when indicating *upper* and *lower* thresholds for that criterion (and in two different rows). For instance, if the threshold value for a criterion related with particles in measuring air quality is 0.04 mg/m^3, it is evident that the "<" symbol should be used, because the total summation has to be as a *maximum,* equal to that value.

If in another example a criterion is considered such as the number of cars with a threshold value of 2,500 cars/hour; then the ">" symbol should be used, since the total summation has to be a *minimum*, equal to that value, so to make sure that the road has the capacity to handle that flow of traffic. If one criterion is cost, we might want to establish that the total amount of the budget is $ 2,390,678 --- no less no more --- and then use the *equal* "=" symbol, since that is the amount of available money.

A criterion can also have a mix of symbols. Say for instance that there is a certain criterion that has a *maximum* sustainable threshold but that can also admit a *minimum.* In that case the "<" symbol is employed to indicate that the selection of alternatives should consider the fact that there is a *maximum* threshold value, and also use, in another row, the symbol ">" to show that this is a *minimum* sustainable value. In other words, it is possible to establish a range or several ranges/s of variation for each criterion.

Similarly, it can be written for the second criterion:

$$a_{21}\, x_1 + a_{22}\, x_2 + a_{23}\, x_3 + a_{24}\, x_4 = b_2$$

And for the **n** criterion:

$$a_{n1}\, x_1 + a_{n2}\, x_2 + a_{n3}\, x_3 + a_{n4}\, x_4 = b_n$$

Therefore, a *set of **n** linear equations* is obtained.

To solve this set of linear equations, that is to say, to find the values for x_1, x_2, x_3 and x_4, we can use the algebra learned in High School, which involved the utilization of the Gauss-Jordan method that implied finding the inverse of a matrix.

Solution

The above set of **n** equations can be expressed in matrix form as:

$$\mathbf{A} \cdot \mathbf{X} = \mathbf{B}$$

If the system is to be solved for "X" a solution is found by determining the inverse of the **A** matrix, and then multiplying it by vector **B**:

$$\mathbf{X} = \mathbf{A}^{-1} \cdot \mathbf{B}$$

However, in the optimization problem one not only wishes to find the "X" values, but also to select *the best alternative or a mix of them,* a process which is called *optimization.*

This is done by using an algorithm named the *Simplex Method* developed by the American mathematician George Dantzig in 1948. Now, in this case a four-dimensional matrix is obtained (the four alternatives), so the problem is stated in a four- dimensional mathematical space. It means that all the linear relationships are in a set of four coordinated axes (remember that we are in a *mathematical space* not in a *physical one*, and then, it is possible to have as many dimensions --- or alternatives, or projects --- as wished, even in the thousands).

The Simplex Method performs many iterations (i.e.; matrix inversions), but each one is represented in a different set of coordinate axes. In order to find the best result, the Simplex Method incorporates a mathematical rule which forces the method to find a better solution in each interaction until it reaches the optimum. This rule is based on the specified objective, which could be: *minimizing environmental costs*, *minimizing total monetary costs, maximizing benefits to the environment*, etc.

Just to better understand this procedure, assume the following elementary situation, with two alternatives which can be then examined in a two-dimensional graphic.

A.3.2 Graphic example
Project: Improvement of an urban highway

Alternatives: (assume just for the sake of the demonstration that it is technically feasible to have a combination of alternatives).

Alternative A: Improve current road by building an additional lane.

Alternative B: Build a local parallel road.

Criteria:

1. Citizens' opinions (unit: number or %)
2. Social benefits (unit: number or %)
3. Air pollution (NOx) (unit: mg/m^3)

Coefficients:

With data from expert opinion, Table A.5 can be built:

Table A.5 **Data**

Criteria		Alternative A	Alternative B	Threshold values
Number	Description			
1	Citizens' opinions	2.60	1.90	2.20
2	Social benefits	0.72	1.14	0.72
3	Air pollution	2.50	2.50	2.05

A set of linear equations can be established as follows:

Citizens' opinion: $2.60\ A + 1.90\ B > 2.20$
Social benefits: $0.72\ A + 1.14\ B > 0.72$
Air quality: $2.50\ A + 2.50\ B < 2.05$

The first row expresses that the contribution of alternative A considering citizens' opinions, plus the contribution of alternative B also considering citizens' opinion, should be *greater* than 2.20 (whatever this number might mean). The second row indicates that the contribution of alternative A considering social benefits plus the contribution of alternative B to the same criterion, must be *greater* than 0.72.

The third row shows that the contribution of alternative A considering the air pollution criterion plus the contribution of alternative B regarding the same criterion, should be *less* than 2.05. Since there are only two alternatives (A and B), they can be represented in a plane with two coordinates axes with A and B as variables. Because there are three criteria expressed as linear equations, they can be drawn as straight lines in this A - B coordinate system.

Figure A.13 shows a graphical representation of criterion 1.

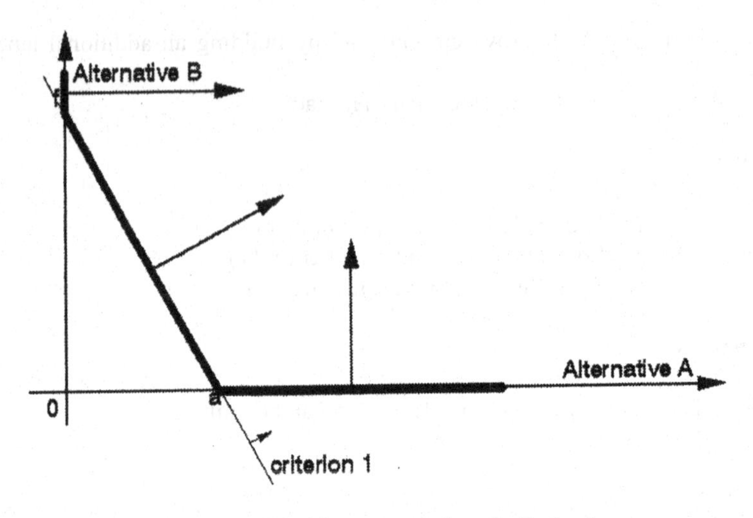

Figure A.13. **Graphic representation of criterion 1**

The little arrow shows the condition of this criterion, i.e., (>) (greater than), and as a consequence the feasible area [axis A - a - f - axis B] will be *above* the line. This area is indicated by the large arrows and is the loci of *all possible solutions for this criterion.*

When the line for criterion 2 is drawn, see Figure A.14, there is also a small arrow that indicates that the solutions for this criterion are *above* the line, as indicated by the ">" sign (greater than), and in the zone limited by [axis A - c - d - axis B],

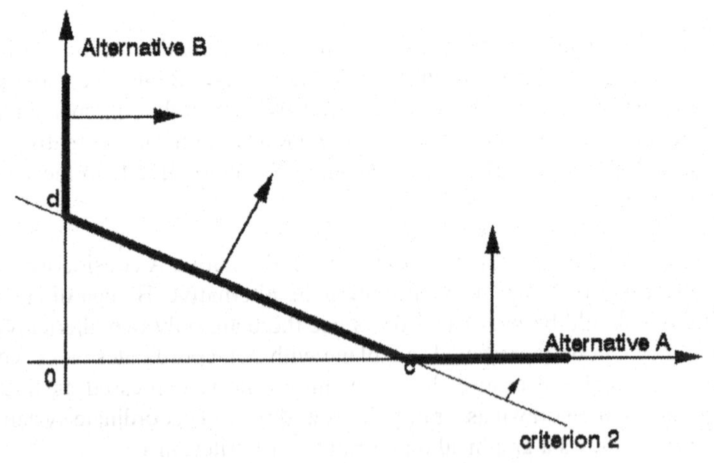

Figure A.14. **Graphic representation of criterion 2**

If the line corresponding to criterion 3 is drawn, see Figure A.15, it has a little arrow opposed to the others, because its sign is "<" (less than).. This indicates that the feasible area for this criterion will be *below* the corresponding line and limited by the coordinates axes A and B. As a consequence this area is now determined by the triangle [e - 0 - b].

So there isa *feasible area* for *each one of the three criteria;* these areas overlap and it can be seen that each one is different from another. Also notice that, in order to include only positive values, these areas have to lie in the first quadrant, that is above axis for alternative A and to the right of axis for alternative B.

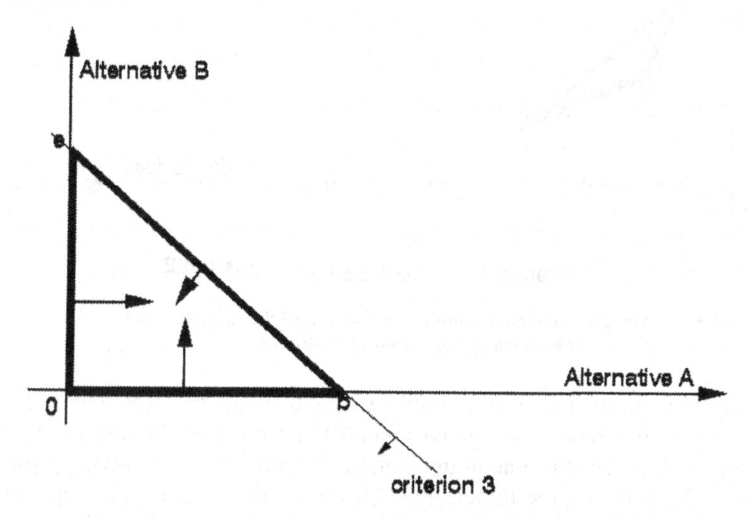

Figure A.15. **Graphic representation of criterion 3**

What it is necessary to do now is to find the *common area to the three criteria.* This area is shown in Figure A.16 as formed by the polygon ***g - h - k***. *Within this polygon lie all the feasible solutions to the problem, and one of the three vertices indicates the optimum solution.*

Assume for instance that this is a minimization problem, that is it has as an objective the minimization of environmental damage. In that case it is easily found that the solution is at the point **h,** and at that point can be found the *degree of participation of each alternative to optimize the objective.*

It can also be seen in this elemental example why MP considers all trade-offs between alternatives and criteria, since *all of them are in the shaded* ***g -h***

- *k* polygon. Notice that points lying out of the shaded area of Figure A.16 do not comply *simultaneously* with the three criteria, and therefore are not optimal solutions, while any point inside that area complies *simultaneously with the three criteria.*

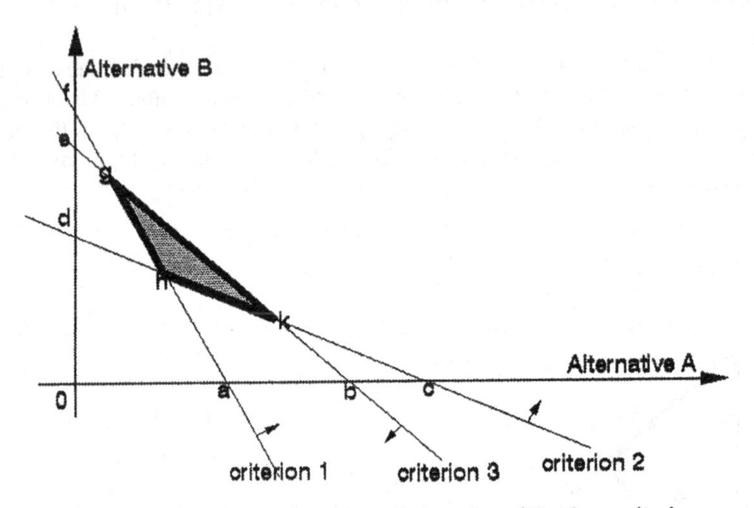

Figure A.16. **Graphic representation of interaction of the three criteria (The linear programming problem)**

In real life problems it is possible to have many projects and many criteria, say 14 projects or alternatives and 10 criteria. This situation cannot be represented unless the maximum number of projects is 3, because our universe is limited to three dimensions. However, this is not a problem for MP since it works with multi-dimensional spaces that we cannot even imagine, let alone see, but finds the right solution using the same reasoning as explained above.

Different results

Decimal values:

There are situations where a mix of alternatives is needed --- whatever their meaning could be --- and then the result must show *decimal values*. For instance we can get as a result that a project for a water treatment plant will produce a flow of 3.8 m^3/sec of water, or that a project for the construction of a highway will imply, in this fiscal year, the construction of 34.25 km.

Integer values:

In other cases fractional values are meaningless. For instance, if a project calls for the construction of houses for low income people, an integer number of houses, such as 256 is needed, but not a value such as 256.67 houses.

The model can be instructed to produce only *integer results*, or even only integer values for some projects and whatever values for others.

Binary values:

In other scenarios, particularly in EIA, just "yes" or " no" answers are needed for projects. In this circumstance the model is instructed to only produce *binary values*, i.e., the result should show a "1" if a project is selected, or a "0" if it is not chosen.

Nonlinear values:

In many circumstances values in the Performance Matrix can be nonlinear. In this event Mathematical Programming uses another algorithm to compute the result, and this is done automatically, so the user doesn't have to specify which algorithm must be utilized. In this last example and in many linear relationships the computer models can utilize more elaborate algorithms such as *branch and bounds* and the *genetic algorithm,* this last one replicating Nature's natural selection of species.

A. 4 Mathematical foundation of input/output analysis

A simple example illustrates the mathematical foundation of the model developed by Wassily Leontief for industrial interrelationships which is nowadays employed all over the world. This is also the model used in Life Cycle Analysis to compute pollution produced in a supply chain (see in Bibliography - Leontief Wassily).

A.4.1 Example - Railway industry

There is interest in determining the economic impact derived from an industrial plant producing passenger and rail freight cars as well as locomotives (railway industry) (RI), to find out how it affects the economy and the productive resources of a country. To simplify this very complex problem just assume that the railway industry receives inputs from only three other sectors of the economy, namely: From the *railway industry itself* (transportation of inputs), from the *steel industry* (production of steel from blast furnaces), and from the *manufacturing sector* (fabrication of parts and assemblies). Table A.6 indicates how much is the total production of the railway industry, expressed in billions of dollars.

Table A.6 **Relationships between the railway industry and its suppliers** (in billions of dollars)

Purchasing sectors ↓	Railway industry	Steel industry	Manufacturing industry	Demand	Total production
Railway industry	1	1	2	6	10
Steel industry	4	2	1	17	24
Manufacturing industry	3	3	6	3	15

Total product of the RI is absorbed in the manner shown. Let us analyze first the *rows* (representing *sales*), of the Table A.6. For the first row:

1 billion dollars go as sales to the RI itself.
1 billion dollars go as sales to the steel industry.
2 billion dollars go as sales to the manufacturing industry.
6 billion dollars are demand from the public.

Similar reasoning can be applied to the other two rows. Let us now analyze the *columns* in Table A.6 which represent *purchases*. As a consequence, for the first column:

1 billion dollars are the purchases by the RI of goods from the RI itself in order to be able to produce a final output of 10 billion dollars, so, 1/10 = **0.1**;
4 billion dollars are the purchases by the RI of goods from the steel industry, therefore 4/10 = **0.4**;
3 billion dollars are the purchases by the RI of goods from the manufacturing industry, hence, 3/10 = **0.3**.

The same reasoning applies to the second and third columns, and then Table A.7 can be obtained.

Table A.7 **Specific consumptions from the railway industry**

Purchasing sectors ↓	Railway industry	Steel industry	Manufacturing industry
Railway industry	**0.1**	0.0416	0.1333
Steel industry	**0.4**	0.0833	0.0666
Manufacturing industry	**0.3**	0.1250	0.4000

The numbers in Table A.7 form the coefficient matrix (A):

$$A = \begin{bmatrix} 0.1 & 0.0416 & 0.1333 \\ 0.4 & 0.0833 & 0.0666 \\ 0.3 & 0.1250 & 0.4000 \end{bmatrix}$$

Now, it is evident that the final demand is, see Figure A.17:

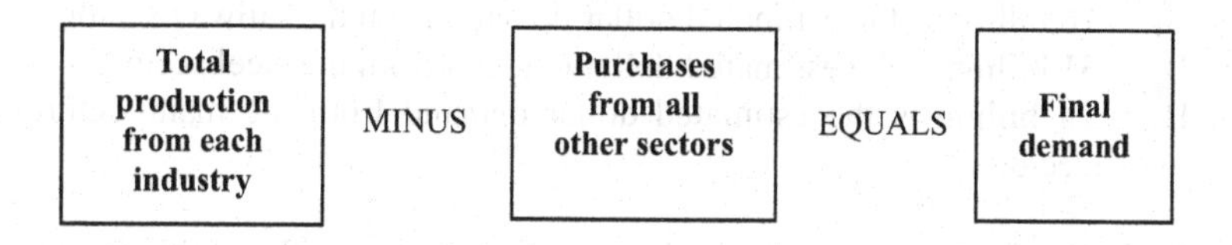

Figure A.17. **Final demand**

If:

X_1 is the total production of the railway industry;
X_2 is the total production of the steel industry;
X_3 is the total production of the manufacturing industry.

Then we will have:

$$\begin{aligned} X_1 - 0.1\,X_1 - 0.0416\,X_2 - 0.1333\,X_3 &= B_1 \\ X_2 - 0.4\,X_1 - 0.0833\,X_2 - 0.0666\,X_3 &= B_2 \\ X_3 - 0.3\,X_1 - 0.1250\,X_2 - 0.4000\,X_3 &= B_3 \end{aligned}$$

Solving this arrangement the following system is obtained:

$$\begin{aligned} (1-a_{11})\,X_1 - a_{12}\,X_2 - a_{13}\,X_3 &= B_1 \\ -a_{21}\,X_1 + (1-a_{22})\,X_2 - a_{23}\,X_3 &= B_2 \\ -a_{31}\,X_1 - a_{32}\,X_2 + (1-a_{33})\,X_3 &= B_3 \end{aligned}$$

But this is the same as taking away the *coefficient matrix* from the *identity matrix* (the matrix equivalent to the unit), which is a matrix with all its elements zero, except in its main diagonal which contains only "1s". As a consequence we get:

$$\text{I - A} = \begin{bmatrix} 0.9 & -0.0416 & -0.1333 \\ -0.4 & 0.9167 & -0.0666 \\ -0.3 & -0.1250 & 0.6000 \end{bmatrix}$$

This is the *Leontief matrix.*

Now, assume that because of market studies, demographic projection and GDP, values have been found for the future demand of goods for each sector or industry, expressed in billions of dollars. Thus:

B_1 = 16 billion is the estimated dollar demand from the railway sector;
B_2 = 31 billion is the estimated dollar demand from the steel sector;
B_3 = 24 billion is the estimated dollar demand from the manufacturing sector.

Since the final demands for each sector have changed, the problem now is to determine, knowing these new demand values, *what are the new industrial interrelationships that will permit the respective productions in each sector to meet these demands.*

This is a problem that has as a solution the following equation:

$$X = (I - A)^{-1} . B$$

Replacing and operating, the result is:

$$X = \begin{bmatrix} 1.2497 & 0.0959 & 0.2882 \\ 0.5988 & 1.1534 & 0.2599 \\ 0.7499 & 0.2883 & 0.8649 \end{bmatrix} \text{x} \begin{bmatrix} 16 \\ 31 \\ 24 \end{bmatrix} = \begin{bmatrix} 29.8849 \\ 51.5738 \\ 65.6933 \end{bmatrix}$$

Therefore, the total output for each sector will, be in billions of dollars:

$$X_1 = 29.88$$
$$X_2 = 51.57$$
$$X_3 = 65.69$$

As can be appreciated these are *cumulative values.* For instance the first row tells us that the *cumulative input from all industries* for a demand of 16 billion dollars from the railway sector will be 29.88 billion dollars. In Life

Cycle Analysis (section 5.6) the same procedure is used, but changing *industrial sectors* to *pollutants*. Therefore, if a dollar value is entered for the production of say 100,000 trucks, the final result will show, considering all the pollution generated by the different sectors to manufacture trucks, the *total cumulative value of contaminants*.

A.5 Statistical notions of Factor Analysis

In Chapter 3, section 3.2. we discussed the use of Factor Analysis (FA) in EIA. In this section a brief and informal presentation is given of the foundations of this important technique. Again, it is explained in plain language without mathematical rigor.
Given a set of variables, FA works with three main techniques:

Variance, a statistical measure of the spread of the data. As a consequence it considers the average distance between a set of points and their mean value.

Correlation matrix, the degree of association between variables, depicted in a matrix.

Linear regression, which allows for the determination of the mathematical relationship between variables.
Taking two correlated variables, the linear regression technique based in the method of least squares, permits the determination of one of the variables when the other is known. Regression cannot detect or establish a cause-and-effect relationship, simply says that the two variables are changing in the same sense, perhaps because there is a common factor, cause or reason for it.

Just to illustrate this point, if in a city a regression analysis shows an increase in people going shopping and at the same time an increase in electricity consumption, probably one fact is not a consequence of the other, albeit perhaps there is a common factor, such as an improvement in the economic situation of its inhabitants, which increments both actions.

In the case of a regression, it is possible to predict the behavior of one variable considering the other, and then one could ask the question: *Why use two variables when one is enough?*

Well, this is exactly what FA analysis does. It takes information from the two variables and packs it into one variable, or in more general terms, it takes information *from a set of many variables* and concentrates it in a *smaller number of variables*, in what are called *factors*. In other words, taking a set of observed variables, FA investigates the relationships within this set of variables, and in so doing it generates another set of theoretical variables called factors.

Factors are new variables which, because of the way they were obtained, *summarize the information from the original variables*. Naturally, there will be some *loss of information*, unless the correlation between both variables is perfect. This loss of information derives from the fact that the diverse points conforming to the cloud of points or scatter plot, are not over a line but dispersed around it, showing a *variance*.

Each original variable has a different degree of contribution to the total variance, so, to *quantify this contribution* we use what is called a *factor loading*, that is also the correlation between each original variable and a factor. Obviously, there is much more about FA, but it is believed that most advanced explanation and documentation can be found in the statistics literature.

A.6 The notion of opportunity cost

This very important concept in economics indicates *the foregone economic development opportunity, when a resource, that is capital, land, labor and time is applied to a specific use, and then is not available for another utilization*. This is something that the human being does many times every day of his/her life: we take decisions about alternative options based on a comparison between the benefits that one option brings against the benefits of another.

Assume for instance that two different projects or alternatives A and B are considered, and both with the same monetary cost, say $ 2,000,000, and a decision has to be made regarding which of them to develop. Suppose now that alternative A will produce a potential benefit of $ 200,000, and alternative B an expected benefit of $ 220,000.

If alternative A is selected, then the opportunity to earn $ 220,000 pursuing alternative B is lost. Conversely, if alternative B is selected then the opportunity to earn $ 200,000 pursuing alternative A is lost. Obviously it is better to develop alternative B, since it has the lower opportunity cost.

However, in this case only the monetary aspect of the transaction is considered. Both alternatives will probably offer different gains and losses related to issues such as contamination, number of people employed, risks, multiplier effect and so on, and as a consequence, each of these issues will also have an opportunity cost.

The *EIA seeks to find an equilibrium between gains and loses offered by the different alternatives*, or, in other words, it looks for a *balance between all the opportunity costs.* There is then a concept of *trade-off values,* and the final consequence is how to play with these trade-off values in order to reach this balance.

Obviously this cannot be made just by comparing alternatives, and for this reason MP offers one of the best solutions to the EIA problem, because this methodology is able to do just that, i.e., to compare all trade-offs between alternatives and criteria in order to optimize a chosen objective/s. Opportunity costs are related to alternatives, therefore in a problem of selecting alternatives, MP indicates how much the optimum decreases when an alternative that is not in the solution is chosen.

A.7 The notion of shadow prices

This is another very important concept in economics. It is related with the constraints or the criteria in an EIA problem, and technically it is the ratio between the derivative of the objective function and the derivative of a constraint. In plain English it means that the shadow price for one criterion indicates *how much the objective function will change* with a *unitary change in the corresponding criterion.*

This is a very useful notion for our purposes because it can be used for *sensitivity analysis*, to investigate what happens when the threshold value of one criterion is relaxed or made tighter. As an example of the first case it could be that one threshold establishes that the *upper limit* of SO_2 discharges is say 135 mg/m^3. What will happen with the solution (the objective), when this upper limit is *relaxed* to 150 mg/m^3? That information is given by the shadow price corresponding to the analyzed criterion.

For a criterion to produce a shadow price it has to be *binding,* that is , *to participate in the solution.* If the solution has some *slack*, then the shadow price is *zero.* This is understandable, since, coming back to our example, if the project produces less contamination than allowed (that is, there is a slack), then it doesn't make any sense to increase this limit.

From that point of view, and because in many cases there are no shadow prices in the solution of a certain MP problem, the *significance of the corresponding criterion is assumed to be proportional to the least slack for each criterion*. Obviously a criterion with slack zero, will have a shadow price and will be probably first in the criteria ranking.

A.8 The notion of sensitivity analysis

In many occasions either in analyzing projects from the economic point of view or from the environmental side, there is a need to investigate the variations in the outcome (the solution) when certain parameters are changed. From the economic standpoint this analysis is important for investigating the effect of variations in prices, operating costs, capital costs, demand, etc. Considering the EIA procedure, many times it is important to determine the influence that a variation in the value of a threshold, such as increasing the water level in a dam to be built, or variations in some upper and lower limits for, say posted speed in a highway, will have in the outcome.

This analysis is called *sensitivity analysis* and is a fundamental component of an EIA Here it is also very important in case of risk and/or uncertainty when the *mean values* are not clearly established or are *unknown.* In these cases upper and lower limits can be imposed on a criterion to let the MP model decide which is the expected value.

Internet references for Appendix

Regression analysis
Title*: Introduction to regression analysis*
Very clear and graphical information about this technique.
http://www.nlreg.com/intro.htm

Correlation analysis
Title: *Technical analysis from A to Z*
Author: Achelis Steven B.
Provides synthetic and good information about this technique.
Eigen analysis
Title: *Eigenvalues and eigenvectors*
Harvey Mudd College
Good information without mathematical jargon.
http://www.math.hmc.edu/calculus/tutorials/eigenstuff/

CONCEPT: Eigenvalues
Title Eigenvalues*: A short introduction*

This is a very clear and well written paper explaining the fundamentals of eigenanalysis and its relationship with the ellipse, and provides a graphical explanation of why this concept is so important to multivariate analysis.
http://obelia.jde.aca.mmu.ac.uk/multivar/eigen.htm

CONCEPT: Circle, ellipse and eigenvectors
Excellent paper where there is a graphical display of the linear transformation. In this case it can be observed how the circle changes into an ellipse. You can play with different correlation values and see how the ellipse changes. See for instance what happens when there are low values of correlation, then improve the correlation, and finally make it a perfect correlation. Check for instance how the ellipse transforms into a straight line for high correlations.
http://www.math.lsa.umich.edu/courses/214/LinearTransf/

Cited references on the Internet by industries

AREA: Waste
Title: *Impact assessment and authorization procedure for installations with major environmental risks.* Environmental Research Programme: Research Area III - --Final report 1999.
Authors: Rabl A., A. Azapagic, C. Blin, B. Burzyinska-Wers, R. Clift, B. Desaigues, S. Dresner, G. Gandara, N. Gilbert, S. Krüger Nielsen, J. Miller, P. Riera, N. Soguel, B. Serensen, J.V. Spadaro, P. van Griethuysen.

Reading of this paper is highly recommended. This extensive (53 pages) report involving six countries in Europe analyses the installation of waste incinerators as well as the construction of landfills. The document is valuable because it is based on the assumption of major environmental risks. It explains ten case studies were a certain procedure was applied to determine the risks involved in each scheme, and with a summary of the environmental impacts of some projects. Particularly interesting is the procedure used to quantify the health risks associated with the source, which is called Impact Pathway Analysis.
This procedure involves:

- Specification of site and technology in the source with quantification of emissions;

- It then applies a model for dispersion of the emissions and, very important, a measure of the concentration of pollutants in the geographical areas affected (receptor sites);
- A dose-response function is then used to find for instance cases of asthma due to the concentration of particulate and gases. Finally, an economic valuation is done This concept is interesting since it considers not only the cost of the treatment for a patient, but also willingness to pay to avoid the suffering.

Each step is thoroughly explained and documented with actual figures from the six countries. A very important issue raised was the necessity to gather public opinion. A questionnaire was prepared with capital questions which are further developed in this paper. It also comments on a contingent evaluation which was performed in two European cities, and explains a cost-benefit analysis of regulations for incinerators, considering thresholds established for emissions and the corresponding costs to meet these thresholds.
http://www-cenerg.ensmp.fr/rabl/pdf/FullReport.pdf

AREA: Dams
Title: *V2 Environmental and social impact assessment for large scale dams*
Authors: Sadler Barry, Iara Verocai and Frank Vanclay.
This extensive paper (48 pages), was prepared by the World Commission on Dams (WCD), a body associated with large dams.
It presents a very interesting approach involving sustainability and types of capital, i.e., the human/social and natural/ecological part of the paper is dedicated to future scenarios on water demands, climate change, and the implications of dams. As an interesting and unusual feature the report details a listing of seven environmentally controversial hydro projects, around the world. Its analysis is interesting because it can help to avoid the mistakes or omissions made in the past, and shows particularly appropriate examples of public participation in Brazil.
http://www.damsreport.org/docs/kbase/thematic/drafts/tr52_draft.pdf

AREA: Agriculture
Title: *Environmental impact guidelines*
Author: FAO Investment Centre, Nov. 1999, Number 1.
Guidelines for formulating projects in agriculture, natural resources, rural development and, poverty alleviation, forestry and fisheries. Offers a case study where the use of EIA contributed to improvements in project design. Details an illustrative outline for environmental assessment reports, as well as

environmental procedures and project timeline of an international financing institution. Details case examples on:

- Rural development in marginal areas programme.
  Mexico: This example shows an environmental assessment for an intermediary lending operation.

- Irrigation rehabilitation.
  Kyrgystan: An example of a sectoral assessment.
  Provides a comprehensive list of references of Banks, government organizations and Bibliography.
  http://www.fao.org/tc/tci/sectors/guideline1.pdf

AREA: Software
Title: *Environmental impact assessment methodologies description and analysis and first approach to environmental impact assessment methodologies application*
Author: The Council of European Professional Informatics Societies -CEPIS.
Indicates available software to download from the Web. Some of the many software are:

* CALINEA4 - Source Dispersion Model.
* CalTOX - A spreadsheet model that relates the concentration of a chemical in soil to the risk of an adverse health effect for a person living or working on or near the contaminated soil.
* ChemSTEER (Chemical Screening Tool for Exposures and Environmental Release).
ChemSTEER estimates occupational inhalation and dermal exposure to a chemical during industrial and commercial manufacturing, processing, and use operations involving the chemical. It also estimates releases of a chemical to air, water, and land that are associated with industrial and commercial manufacturing, processing, and use of the chemical.
* Comparative Risk Assessment - By US Environmental Protection Agency and Purdue University. Contains the history and methodology of comparative risk, as well as many case studies and information sources. It is intended for those involved in government, academia, public health, public interest groups and local communities.
http://www.cepis.ops-oms.org/muwww/fulltext/repind51/environ/environ.html

AREA: Transportation
Title: *Capital beltway corridor --- Transportation study*
Evaluation factors.
http://www.rkkengineers.com/sha/capital/evalfactors.htm

AREA: Transportation
Title:*OECD/ECMT Conference on strategic environmental assessment for transport*
Author: Marian Tracz
Strategic environmental assessment for transport in Poland.
http://www1.oecd.org/cem/topics/env/SEA99/SEAtracz.pdf

AREA: Transportation
Title: *Capital beltway corridor --- Transportation study*
It is a summary of a transportation study for a corridor in the State of Maryland, USA.
In an environmental summary, provides information about an environmental inventory to identify socio-economic, cultural and natural environmental resources. As in many studies the status-quo condition (to do nothing) was considered. It is interesting to notice that the area includes 136 parks, with a varied land use, and also incorporates many historic sites. The study also identifies 12 stream crossings and recognizes that wetlands and floodplains could be affected. It also shows an impact qualitative matrix with three different types of potential impacts,--- minor, moderate and major ---and with seven alternatives evaluated through two main criteria which are: socio-economic environmental and natural environment. The first one is subdivided in four sub-criteria and the latter in seven criteria.
http://www.rkkengineers.com/sha/capital/environmentalsummary.htm

AREA: Transportation
Title: *Strategic environmental assessment of transport corridor: Lessons learnt comparing methods of five Member States*
Author: Olivia Bina.
This is a publication of the European Commission. It is the analysis of five corridors in different countries. Provides comprehensive information on:

- Scoping;
- Different types of analysis;
- Manner in which each country faces a problem.

http://europa.eu.int/comm/environment/eia/sea-studies-and-reports/sea_transport2.pdf

AREA : Transportation

Title*: Template 10: Cost-Effectiveness worksheet --- incremental cost per incremental*

Published by FTA-Federal Transit Administration, United States Department of Transportation.

http://www.fta.dot.gov/library/policy/ns/2002/tem10.html

AREA: Airports

Title: *Summary environmental impact assessment of the New Samarinda Airport --- Eastern Islands Air Transport Development Project*

Published by: Asian Development Bank.

It gives information on:

- Loss of ecological, cultural or other resources;
- Environmental impacts due to project design;
- Environmental impacts during construction;
- Impacts during operation;
- Alternatives;
- Cost-Benefit analysis - Internal Rate of Return;
- Economics benefits;
- Project costs;
- Monitoring and reporting costs;
- No quantified environmental impacts;
- Institutional capability;
- Monitoring program;
- Public involvement.

http://www.adb.org/Documents/Environment/Ino/ino-samarinda-airport.pdf

AREA: Airports

Title: *Draft environmental impact statement/environmental report (Draft EIS/EIR)*

Comprehensive information about EIA for the Santa Barbara Airport (USA). with detailed information about impacts and their significance with a particular emphasis on wildlife and land use.

http://www.flysba.com/pdf/EIS_EIR/summary.pdf

AREA: Navigation

Title*: Report on environmental impact assessment: The navigation channel improvement project of the Lacang-Mekong River from China-Myanmar boundary, marker 243, to Ban Houei Sai of Laos*

http://www.irn.org/programs/mekong/021018.critiquefisheries.pdf

AREA: Coal
Title: *Environmental impact statement --- Moorvale coal project*
Published by: MACARTHUR COAL
Extensive analysis on:

* Environmental values affected;
* Management of environmental values;
* Management of greenhouse gas emissions.

http://www.macarthurcoal.com.au/EISFeb2002/EISFeb2002-Part4.html

AREA: Environment
Title: *Sustainable technology system analysis*
Published by: U.S. Environmental Protection Agency.
This publication involves several issues with all of them having a fundamental importance in environmental management. They are:

- Simulation and design: Involves models for pollution reduction in the chemical industry;
- Environmental engineering economics;
- Life Cycle Assessment;
- Impact assessment and measurement;
- Engineering trade-offs (ETO);
- Total integration for sustainable development.

http://epa.gov/ORD/NRMRL/std/sab/ETO_CONCEPT.htm

AREA: Oil
Title: *Criteria for classifying the level on environmental impact of regulated activities: Requirement under Part 12 Petroleum Act 2000*
Published by: Petroleum Group - Department of Primary Industries and Resources (South Australia) - PIRSA.
Very interesting paper which, amongst other subjects details very good examples of:

- Criteria for determining level of environmental impact;
- Predictability criterion;
- Manageability criterion;
- Environmental significance against manageability criterion;
- Environmental significance;
- Classifying activity environmental impact;
- Examples of events and their environmental consequences.

http://www.pir.sa.gov.au/pages/petrol/environmental_reg/documents/criteria_guidelines.pdf

AREA: Oil industry
Title: *Partnering and Environmental Impact Assessment*
Case study presented by Shell Petroleum Development Company to revitalize its Environmental Impact Assessment Procedures.
http://www.bpd-naturalresources.org/media/pdf/spdc/spdc_eia.pdf

AREA: Civil urban construction
Title: *Construction and city related sustainability indicators*
Published by CRISP Sustainable Construction - European Union.
As per their own words: *"CRISP aims to develop and validate harmonized criteria and relevant and efficient indicators to measure the sustainability of construction projects particularly within the urban built environment".*
http://crisp.cstb.fr/default.htm

AREA: Civil urban Construction
Title: *Sustainability profile for a location (DPL) - The Netherlands*
Author: R.W. Lanting. - CRISP: Construction and City Related Sustainability Indicators. This is a publication of the European Union. This is an example in The Netherlands and as per their own words:

"The objective of the project is to develop a methodology to assess the sustainability of new or refurbished areas (district level). The tool is to be used in the planning phase for comparing different development plans for their sustainability. Indicators are used to assess ecological, social and economic sustainability. The result is displayed as a bar diagram (profile) The first prototype has been tested in 3 cases (refurbishment plans for 3 urban districts in 3 municipalities). Ongoing research concerns the fine tuning of the indicators".
http://crisp.cstb.fr/view_rdworks.asp?id_rdworks=12

AREA: Water
Title: *Case studies*
Report prepared by the Canadian International Development Agency (CIDA) on three cases studies. It is interesting the use of the matrix for environmental issues, for the three projects.
http://www.acdi-cida.gc.ca/ea/studies

AREA: Water

Title: *Economic Analysis of Designating Outstanding National Resource Waters in Tennessee: Theory and an Application in Monroe County*

Author: Paul Jaus
Comment on the use of Cost-Benefit analysis.
http://web.utk.edu/~casnr/spot_jakus.htm

AREA: Electricity
Title: *Uncertainty and the Cost-Effectiveness of regional NOx emissions reductions from electricity generation*
Authors: Dallas Burtraw, Ranjit Bharvirkar, Meghan McGuiness.
Comprehensive report (55 pages) over NOx emissions due to electric generation, and the use of Cost-Effectiveness Analysis.
http://www.rff.org/multipollutant/disc_papers/PDF_files/0201.pdf

ROAD MAP FOR ENVIRONMENTAL IMPACT ASSESSMENT

Flowchart and road map for Environmental Impact Assessment

This section contains a road map that you might want to use for your EIA.

In the next pages you will find a flowchart with its corresponding explanation. The flowchart succinctly shows 12 general stages for this task. Each stage is indicated in a rectangular box with a number to reveal sequence, which is also depicted by arrows. So, a stage cannot commence until the precedent task or tasks have been completed.

You can in your own project establish duration estimates, in days, weeks or months (use the small square box in the upper right angle of the stage box), to complete each independent task, but using the same kind of units in all boxes. This way you will be able to calculate the total period of time needed to get a complete EIA report done just by adding up the durations that you have established.

Observe that some stages are not in series but can be executed simultaneously, in parallel with others, as in stages 5, 6 and 7. Most probably they will take different durations to be completed, say for instance **three weeks** for stage 5 **two weeks** for stage 6 and **six weeks** for stage 7. According to the flowchart you need to have stages 5, 6 and 7 finished to commence stage 8. Assuming just as an example, some durations for stages 1, 2, 3 and 4 it is obvious that you can start stage 8 only after say:

3 weeks, to complete stage 1, plus:
2 weeks to complete stage 2, and:
4 weeks to complete stage 3. Notice that stages 2 and 3 are in parallel, however we cannot start stage 4 if the two of them have not finished, that is we have to consider the largest duration, which is 4 weeks. Plus:
6 weeks to complete stage 4, plus:
6 weeks to complete stage 7 (the three weeks deemed necessary for execution of stage 5 as well as the two weeks needed for stage 6 are irrelevant since you need six weeks for stage 7 in order to be able to start stage 8).

Therefore, you can start stage 8 after: 3+4+6+6 = 19 weeks. This way, and continuing up to the end, the whole duration of the process can be estimated, and you will be able to answer the question that surely will be posed to you as a person responsible for this task:

How long do you think it will take?

The pages following this flowchart constitute a working sheet in an expanded flowchart format and that you can use for the project you are preparing or analyzing.

In almost each topic it is indicated between square brackets the section of the book from where you can get information about a specific subject.

Needless to say, sometimes and because many circumstances, events do not evolve in the indicated manner and some back and forth must be expected, but in general, the flowchart follows the logical sequence of events.

Explanation of the flow chart:

Stage 1: From feasibility studies the *type of project* is known.
Stage 2: The type of project indicates the *geographical areas* it affects.
Stage 3: From the feasibility study the different *alternatives or options* for a project are known.
Stage 4: *Impacts* are determined.
Stages 5, 6 and 7: Criteria *determination*, *weights* and *thresholds* are identified here.
Stage 8: *Information* for the EIA is gathered here.
Stage 9: The *performance matrix* is developed.
Stage 10: Takes place the *selection of projects or options*.
Stage 11: The *whole plan is revised* and perhaps modified after consultation with different actors.
Stage 12: The final *EIS report* is written

FLOWCHART FOR ENVIRONMENTAL IMPACT ASSESSMENT

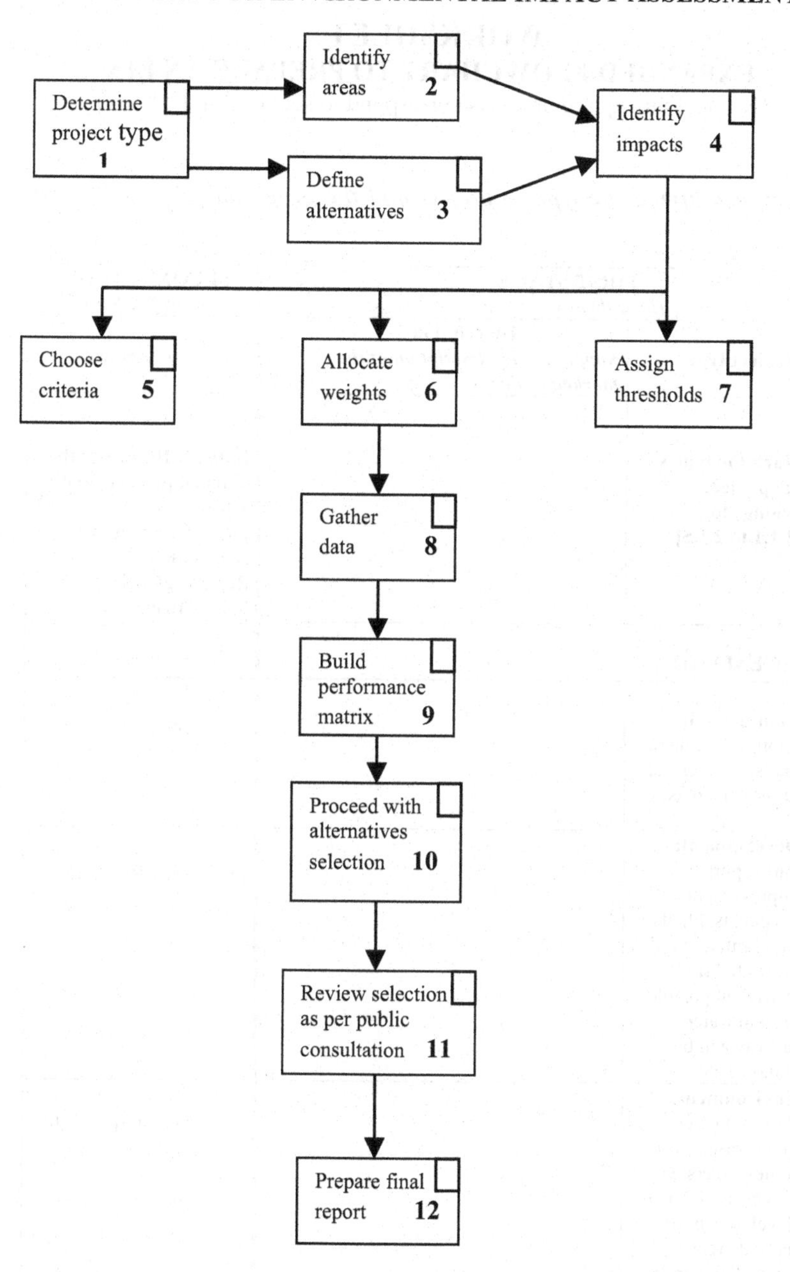

WORKSHEET

EXPANDED FLOWCHART TO PREPARE AN EIA

(Numbers in square brackets correspond to sections in the book)

Start by identifying the type of project and the geographical area/s it will affect

FIRST STAGE *(Identify project type)*			SECOND STAGE *(Identify area/s)*
Project type		**Description** ***Supply a brief description of the marked project/s***	**Areas affected**
	Mark the type/s the project belongs to [2.1.1 to 2.1.5]		**Indicate the area/s the marked project/s will affect** Int.: International Nat. National Reg.: Regional Urb.: Urban
	GENERAL		
☐	**Commercial:** Shopping centers, malls, recovery of run-down areas		☐ Int. ☐ Nat. ☐ Reg. ☐ Urb.
☐	**Development:** Parks, ports improvement, aqueducts, plants for elimination of unwanted sub-stances in potable ground water, railway and bus stations, etc.		☐ Int. ☐ Nat. ☐ Reg. ☐ Urb.
☐	**Environment:** Parks construction, green spaces, bike routes, rivers, lakes and ports clean up, development of natural areas in cities, large under-ground water reservoirs, etc.		☐ Int. ☐ Nat. ☐ Reg. ☐ Urb.

☐	**Industrial:** Any type of industrial activity		☐ Int. ☐ Nat. ☐ Reg. ☐ Urb.
☐	**Social:** Construction of sports and social clubs, recreation areas		☐ Int. ☐ Nat. ☐ Reg. ☐ Urb.
☐	**Infrastructure:** Roads, storm-drains, urban intersections, urban transportation		☐ Int. ☐ Nat. ☐ Reg. ☐ Urb.
	SPECIFIC		
☐	**Agriculture:** Large schemes and new plantations, construction of drainage ditches		☐ Int. ☐ Nat. ☐ Reg. ☐ Urb.
☐	**Airports:** Terminals, run-ways, utilities and access roads		☐ Int. ☐ Nat. ☐ Reg. ☐ Urb.
☐	**Chemicals:** Production and transport		☐ Int. ☐ Nat. ☐ Reg. ☐ Urb.
☐	**Communications:** Antennas, switching equipment, underground works, etc.		☐ Int. ☐ Nat. ☐ Reg. ☐ Urb.
☐	**Culture**, includes large projects such as theaters, muse-ums, amphithea-ters, etc.		☐ Int. ☐ Nat. ☐ Reg. ☐ Urb.

☐	**Electronics:** Includes chips production, high-tech equipment, computers, radar, sonar, TV, etc.		☐ Int. ☐ Nat. ☐ Reg. ☐ Urb.
☐	**Energy:** Generation and distribution, as well as the use of non-conventional sources		☐ Int. ☐ Nat. ☐ Reg. ☐ Urb.
☐	**Entertainment** includes thematic parks, as well as open-air cinemas.		☐ Int. ☐ Nat. ☐ Reg. ☐ Urb.
☐	**Food industry:** Includes industrial plants and meatpacking plants		☐ Int. ☐ Nat. ☐ Reg. ☐ Urb.
☐	**Forestry:** Exploitation, reforestation and timber processing. Includes necessary service roads		☐ Int. ☐ Nat. ☐ Reg. ☐ Urb.
☐	**Housing,** applies to large developments, including utilities		☐ Int. ☐ Nat. ☐ Reg. ☐ Urb.
☐	**Hydro,** including dams, lakes and energy generation and transportation)		☐ Int. ☐ Nat. ☐ Reg. ☐ Urb.
☐	**Land use**: Natural parks and reserves, golf courses, change of land use		☐ Int. ☐ Nat. ☐ Reg. ☐ Urb.
☐	**Large farms:** Cattle, pig and poultry operations, manure treatment plants		☐ Int. ☐ Nat. ☐ Reg. ☐ Urb.

☐	**Large irrigation** schemes including irrigation dams		☐ Int. ☐ Nat. ☐ Reg. ☐ Urb.
☐	**Mining:** Open pit and underground mines, service roads, minerals disposal, ore processing, metal refining and treatment plants		☐ Int. ☐ Nat. ☐ Reg. ☐ Urb.
☐	**Navigation:** Channels dredging, construction of container handling facilities for river and sea ports		☐ Int. ☐ Nat. ☐ Reg. ☐ Urb.
☐	**Oil and gas:** Production, transportation, refining. Includes construction and operation of oil and gas pipelines		☐ Int. ☐ Nat. ☐ Reg. ☐ Urb.
☐	**Pharmaceuticals:** Drugs, antibiotics		☐ Int. ☐ Nat. ☐ Reg. ☐ Urb.
☐	**Public health:** Hospitals		☐ Int. ☐ Nat. ☐ Reg. ☐ Urb.
☐	**Recreation:** Thematic parks, amusement parks		☐ Int. ☐ Nat. ☐ Reg. ☐ Urb.
☐	**Scientific**, including construction of linear accelerators, tomahawks, etc.		☐ Int. ☐ Nat. ☐ Reg. ☐ Urb.

☐	**Sea resources**: Sea creatures exploitation and processing, as well as exploitation of mineral resources such as chemicals, manganese, etc.		☐ Int. ☐ Nat. ☐ Reg. ☐ Urb.
☐	**Sewage:** Sewerage networks, treatment plants		☐ Int. ☐ Nat. ☐ Reg. ☐ Urb.
☐	**Shipyards:** Naval construction, marine platforms, cranes, etc.		☐ Int. ☐ Nat. ☐ Reg. ☐ Urb.
☐	**Sports,** includes modification of land usage as in large structures, stadiums, racetracks and golf courses		☐ Int. ☐ Nat. ☐ Reg. ☐ Urb.
☐	**Tourism:** Construction of large hotels and tourism centers. Roads and facilities improvement. Construction of walking paths in forests, etc. Construction of mountain sports facilities		☐ Int. ☐ Nat. ☐ Reg. ☐ Urb.
☐	**Transportation:** Construction of roads, railroads, subways, elevated trains, magnetic levitation, water channels, etc		☐ Int. ☐ Nat. ☐ Reg. ☐ Urb.
☐	**Urban:** Downtown rehabilitation plans, relocations, river banks consolidation, modification of water courses,		☐ Int. ☐ Nat. ☐ Reg. ☐ Urb.

	construction of bridges, tunnels, and urban intersections, urban river channeling		
☐	**Wastes** (dangerous): Transportation and disposal		☐ Int. ☐ Nat. ☐ Reg. ☐ Urb.
☐	**Wastes** (domestic): Construction of landfills, incinerators, recycling plants		☐ Int. ☐ Nat. ☐ Reg. ☐ Urb.
☐	**Wastes** (industrial): Construction of disposal sites		☐ Int. ☐ Nat. ☐ Reg. ☐ Urb.
☐	**Water:** Water catchments works, treatment and distribution		☐ Int. ☐ Nat. ☐ Reg. ☐ Urb.
☐			☐ Int. ☐ Nat. ☐ Reg. ☐ Urb.
☐			☐ Int. ☐ Nat. ☐ Reg. ☐ Urb.
☐			☐ Int. ☐ Nat. ☐ Reg. ☐ Urb.

GO to <u>THIRD STAGE</u>

And define the alternatives or options for the project (from the feasibility study) **[2.1.1] [2.1.2] [2.1.3] [2.1.4]**

Are there a-priori weights assigned to the alternatives? **[2.1.3]**

□ **YES** □ **NO**

If **YES,** ***explain on what basis those a-priori weights have been allocated***

Alter-native or option	**Brief description of each alternative or option**	**Main features**	**Cost**	**A priori weight**
A				
B				
C				

D				
E				

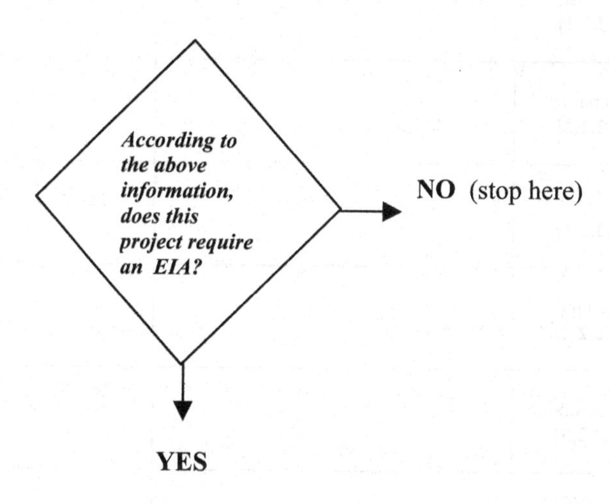
According to the above information, does this project require an EIA?
NO (stop here)
YES

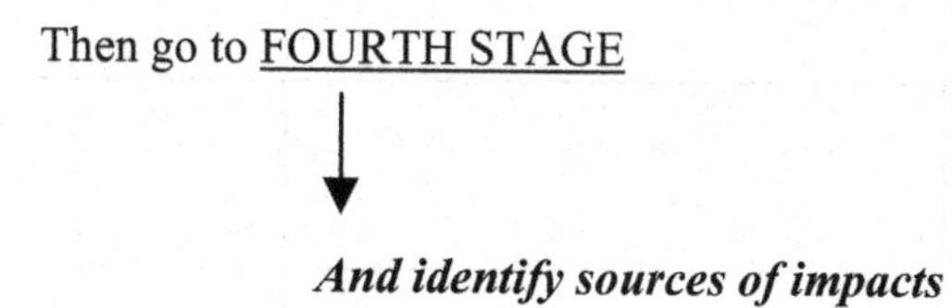
Then go to FOURTH STAGE
And identify sources of impacts

Are these impacts

- ☐ Generated by the project itself?
- ☐ From backwards activities? **[3.3.2]**
- ☐ From forwards activities? **[3.3.2]**
- ☐ Others

See also as a reference **Internet references for Chapter 2**

Determine potential impacts caused by the project

	Impacts	Briefly detail reasons for identifying these impacts	Expected mitigation [2.2.6] Explain
☐	Positive **[2.2.2.1]**		
☐	Adverse **[2.2.2.1]**		
☐	Primary **[2.2.2.2]**		
☐	Secondary **[2.2.2.2]**		
☐	Tertiary **[2.2.2.2]**		
☐	Measurable **[2.2.2.3]**		
☐	Indetermi-nate **[2.2.2.3]**		
☐	Apparent **[2.2.2.4]**		
☐	Cumulative **[2.2.2.5]**		
☐	Residual impact **[2.2.2.7]**		

☐	Spatially related **[2.2.2.8]**		
☐	Temporal **[2.2.2.9]**		
☐	Reversible **[2.2.2.10]**		
☐	Likelihood **[2.2.2.11]**		
☐	Unexpected **[2.2.2.12]**		
☐	Others		
☐	Risk effects **[2.2.2.13]**		
☐	Residual effects **[2.2.2.14]**		
☐	Population **[2.2.2.15]**		
☐	Interaction between impacts **[2.2.2.16]**		

Go to <u>FIFTH STAGE</u>

↓

And define areas to be covered by criteria

Choose criteria **[3.1, 3.1.1 to 3.1.9]**

What type of criteria are considered in this study?

			Where this information comes from			
Mark type of criteria of choice		**On what basis is each one chosen?**	**Tech-nical studies and reports**	**Similar projects**	**Sur-veys**	**Sta-tis-tics**
☐	Technical **[3.1.1]**					
☐	Environment **[3.1.2]**					
☐	Safety **[3.1.3]**					
☐	Social **[3.1.4]**					
☐	Economic **[3.1.5]**					
☐	Construction **[3.1.6]**					
☐	Spatial **[3.1.7]**					
☐	Temporal **[3.1.8]**					

☐	Significance **[3.1.14] and chapter 3 Internet references**					

What source do you use for criteria selection?

☐ Expert opinion **[3.1.10]**
☐ Political factors **[5.9.1] [5.11] [6.4]**
☐ Citizens' opinion **[3.3.6]**
☐ Other

Do you think that criteria in this project have to be independent from each other? **[3.1.15] [3.2]**

☐ **YES** ☐ **NO**

If **YES** ***what measures are being taken to assure independency?***

Who analyses criteria?

☐ Technical people? **[3.3.6]**
☐ Stakeholders? **[2.1.5.2] [3.3.6] [5.7] [5.9]**
☐ Citizens? **[3.3.6]**
☐ Other?

Go to SIXTH STAGE

↓

Regarding criteria weights:

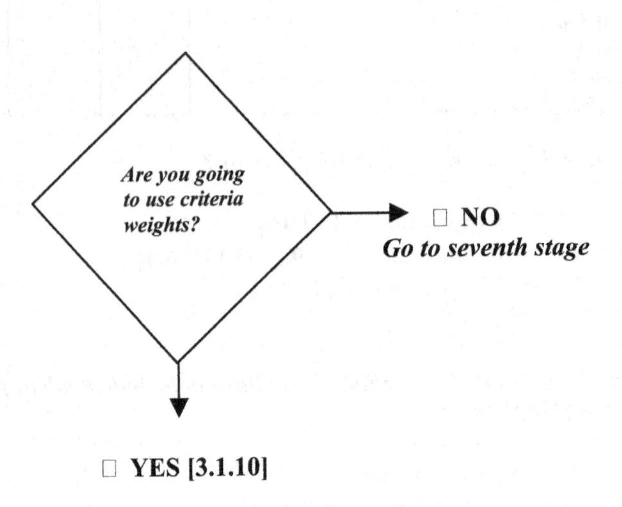

Then, allocate weights to criteria **[3.1.10]**

Criteria	Description of criteria	Weights
Criterion 1		
Criterion 2		
Criterion 3		
Criterion 4		
Criterion 5		
Criterion 6		

Mark in the following boxes who assigned weights

- ☐ Expert opinion **[3.1.10]**
- ☐ Political factors **[5.9.1] [5.11] [6.4]**
- ☐ Citizens' opinion **[3.3.6]**
- ☐ Other

Go to SEVENTH STAGE

↓

***And assign thresholds to criteria* [3.1.11 to 3.1.12]**. See also **Internet references for Chapter 3**

Indicate on what basis have thresholds values been selected?

- ☐ Based on international, national or local standards and indicators **[6.6]**
- ☐ Threshold standards **[3.1.12]**
- ☐ Using technical indicators
- ☐ People's preferences **[5.3] [5.3.2] [5.8][5.8.1]**
- ☐ Minimum value in the corresponding performance matrix row **[6.8.1]**
- ☐ Resource availability **[1.7] [1.10] [6.4]**
- ☐ Low and high values for a certain criterion **[5.7.3.1 under Thresholds]**
- ☐ Values for criteria involving risk **[1.13] [2.2.2.13] [A.8]**
- ☐ Other

Go to EIGHTH STAGE

↓

And specify which are the techniques used to gather data for this project

	Tools *Mark tools used*	Format *Indicate format*	Give details on reasons for selection	Background information *Specify source of information to justify marks and values.*
☐	Checklists **[4.3.1]**			
☐	Network analysis **[4.3.2]**			
☐	Leopold's matrix **[4.3.3]**			
☐	McHarg system **[4.3.6]**			
☐	Delphi method **[4.3.7]**			
☐	Dose-response function **[4.3.8]**			
☐	Stepped-matrices **[4.3.9]**			

What studies have been done to further support this selection?

Go to NINTH STAGE

And build a performance matrix similar to the matrix depicted in next page **[2.1.4] [2.1.5.1] [5.7.1.1] [5.8.1] 5.9.1] [A.31]**

Performance matrix

		Alternatives			
		A	**B**	**C**	**Thresholds**
	Cost→				
	Weight ↓				
Criterion 1					
Criterion 2					
Criterion 3					
Criterion n					

Indicate where the data for the performance matrix comes from

- ☐ Expert opinion **[3.1.10]**
- ☐ Political factors **[5.9.1] [5.11] [6.4]**
- ☐ Citizens' opinion **[3.3.6]**
- ☐ Technical people
- ☐ Stakeholders **[2.1.5.2] [3.3.6] [5.7][5.9]**
- ☐ People. In this case indicate how the information has been obtained.
  - ☐ Surveys (in this case show results)
  - ☐ Meetings (make a brief description and number of meetings held)
- ☐ Do the values for the different alternatives reflect the fact that some alternatives are exclusive (that is they stand alone), that some must complement each other, and that there could also exist a sequence? **[1.2] [2.1.1] [6.2]**

Go to <u>TENTH STAGE</u>

And proceed with alternatives selection

	Tool/s used for selection	**Briefly explain the fundamentals for using a particular tool or several tools**	**Assumptions made**
☐	GIS [5.1]		
☐	CV [5.2]		
☐	CBA [5.3]		
☐	CEA [5.4]		
☐	IO [5.5]		
☐	LCA [5.6]		
☐	AHP [5.8]		
☐	MP [5.9]		

What was the result of the selection?

Were you looking for the selection of just an alternative or for a ranking of them?

Go to <u>ELEVENTH STAGE</u>

↓

Review all results obtained and get feed-back from consultation

Indicate who intervened in this consultation process?

- ☐ Expert opinion **[3.1.10]**
- ☐ Political factors **[5.9.1] [5.11] [6.4]**
- ☐ Citizens' opinion **[3.3.6]**
- ☐ Technical people
- ☐ Stakeholders **[2.1.5.2] [3.3.6] [5.7][5.9]**

Are sensitivity analyses being performed? **[4.3.6] [5.9 under Comments on results] [5.11] [A.8]**

☐ **YES** ☐ **NO**

If **YES** ***which parameters were used and what are the conclusions?***

Go to <u>TWELFTH STAGE</u>

↓

And prepare your final EIA report **[1.15]**

GLOSSARY

On EIA

Appraisal: The act of estimating or judging the nature or value of a project.
Assessment: To estimate officially the value of an action, for instance, assessment of the damage done by a specific impact.
Estimate: To render an approximate judgment or opinion regarding the value of something.
Evaluate: To establish the value or worth of something, for instance the significance of an impact can be evaluated.
Evaluation: The act of evaluating.
Expert opinion or expert judgment: Opinion or judgment by persons knowledgeable on some subject.
Decision makers: Persons making decisions about the convenience or not of pursuing certain projects or alternatives.
Disposable income: Economic concept that expresses the total amount of income that is available to a household for consumption or savings over a certain period of time, after taxes and other fiscal obligations have been met. This is different from total income, since the latter is before taxes and obligations. Total input includes some other payments such as unemployment insurance.
EU: European Union.

On Environment

Bhopal: City of India, place of a human disaster due to release of chemicals.
BOD_5: Biochemical Oxygen Demand - A measure of oxygen in water.
Complementary projects: Projects for which the execution of one of them also involves the execution of the other.
Ecology: The science that investigates the relationship between organisms and their environments.
City footprint: The area involving a city and the surrounding area from where the city extracts resources.
Landfill: A piece of land devoted to the dumping of waste.
Linear transformation: A mathematical function than operates between vector spaces. Given a source vector space it finds the correspondance of that vector in another vector space, using additive and scalar multiplication operations.
CO_2 sinks: Method to compensate the emission of CO_2 with the development of gas absorbing trees.
Distributional impact: How impacts affect different parts of the environment.
Exclusive projects: When the execution of one project precludes the execution of another.

FAO: Food and Agriculture Organization of the United Nations.
Floatation: Mechanical process used to separate minerals, employing pulverized ore, water and chemicals.
Fuzzy: A mathematical system where the values of variables are not very well defined.
GDP: Gross Domestic Product. System to measure economic development.
GIS: Geographic Information System, involving georeferenced data.
Global Urban Observatory: System developed by United Nations to keep track of advances in the use of urban indicators.
Global warming potential (GWP): The potential of a component to increase global warming.
Impact: Can be defined as the change of some conditions in human health and in the ecosystem caused by the development and implementation of a project.
Indicators: Values or metrics established to measure some issues.
Information Theory (IT): Branch of Mathematics using Statistics, founded by Claude Shannon (see Bibliography - Shannon C.E.). It deals with aspects of communications related with the measurement of the quantity of information that can be transmmited through channels and the efficiency of the transmission process.
Local Urban Observatory: A Global Urban Observatory at local level.
LPG: Liquefied Propane Gas.
Matrix: A table or arrangement of numbers.
Marginal value: The determination of how the final solution changes with a unit change in the criterion value.
Model: In the case of a mathematical model it is a representation of an actual system.
Multiplier effect: A certain project can generate an increase in the spending habits of the population it affects. This spending produces in turn more spending, which also creates further expenditures. The relationship between the total spending and the original one is called "multiplier effect".
See Vivian C. Choi's brief article with a very clear explanation at: http://www.admin.gov.gu/commerce/multiplier.htm
Different multiplier effects can be calculated for such activities as tourism, construction of large industrial complexes, etc.
NEPA: National Environmental Policy Act. A key piece of legislation in the USA.
NOx: Oxides of nitrogen.
Pair-wise matrix: Given a set of criteria, a pair-wise comparison evaluates one criterion with each one of the others in the form of a matrix.
Particulate matter: Solid particles of matter found in atmospheric air.
Radon: Radioactive gas produced by the decay of radium products.
Remediation: Measures taken to restore initial conditions or to ameliorate a damage.

Restoration: Works executed to restore something to its original condition.
Risk: Can be defined as the existing possibility of some sort of damage occurring in the future as a consequence of the project's actions. The risk involves both human health and ecosystems alike.
Scoring: Establishing values for the relationships between alternatives.
Sensitivity analysis: Testing the resulting output of a model when certain parameters are changed.
Significance: Importance of the residual impact after remediation measures.
Stakeholders: People who may be affected by the impact of a project.
Supply chain: The chain or sequence of inputs needed to manufacture a product or to develop a project.
Tailings: Liquor from mining and refining operations.
TSP: Total suspension particulate in the air.
Uncertainty: When it is impossible to calculate the results from project's actions.
United Nations Habitat: United Nations Agency dealing with housing.
VOC: Volatile organic compounds. These are a large list of substances that can react photochemically in the atmosphere.
Weighting: Assigning a degree of importance or establishing a hierarchy to projects or alternatives and criteria.

On Sustainability

Sustainable development: A development that attends the necessity of the present population but keeping resources to be also used by future generations.
Sustainable indicators: Indicators that take into account sustainability.

BIBLIOGRAPHY

Alberti, M. et al.: 1996, "Measuring urban sustainability" - Special issue - *Environmental Impact Assessment Review*, **16**.

Atkisson, Alan: 1996, "Developing indicators of sustainable community: Lessons from sustainable Seattle". *Environmental Impact Assessment Review*, **16**, 337--350.

Bisset, R.: 1996, "UNEP. EIA Training Resource Manual - EIA: Issues, Trends and Practice", *United Nations Environmental Programme (UNE) - Economic and Trade Unit Division of Technology, Industry and Economics (UNEP)*, Geneva, Switzerland. http://www.cenn.org/Books/bisset/

Bond, A. and Stewart, G: 2002, "Environmental agency scoping guidance on the environmental impact assessment of projects". *Impact Assessment and Project Appraisal* , **20**, 135--142- Beech Tree Publishing, Surrey, U.K.

Crookes, W. and De Wit, M.: 2002, "Environmental economic valuation and its application in environmental assessment: an evaluation of the status quo with reference to South Africa". - *Impact Assessment and Project Appraisal*, **20,** 127--134 - Beech Tree Publishing, Surrey, U.K.

Dahdouh-Guebas, F., Triest, L., and Verneirt, M.: 1998, " The importance of a hierarchical ecosystem classification for the biological evaluation and selection of least valuable sites", *Impact Assessment and Project Appraisal*, **16**, 185--193- Beech Tree Publishing, Surrey, U.K.

Dantzig, G.B.: 1951, *Maximization of a linear function of variables subject to linear inequalities - Activities analysis of production ands allocation.* Pages 339-347, John Wiley & Sons, Inc. New York, USA.

Dey, P.K: 2001, "Integrated approach to project feasibility analysis: a case study". *Impact Assessment and Project Appraisal.* **19,** 235--245 - Beech Tree Publishing, Surrey, U.K.

Eggenberger, N.M. and Partidario, M.R.: 2000, "Development of a framework to assist the integration of environmental, social and economic issues in spatial planning". *Proceedings of the 19th IAIA99 Conference,*

Glasgow, Scotia, - International Association for Impact Assessment, Fargo, ND, USA.

Environment Canada: 1995, "Measuring urban sustainability: Canadian indicators workshop" - Workshop proceedings Toronto, Canada.

Environment Canada, 1995. "Municipal State of the Environment Reporting in Canada: Current Status and Future Needs". *Occasional Paper No. 6, State of the Environment Directorate*, Ottawa. Table 13 – Most common municipal indicators. Authors: Monica Campbell and Virginia Maclaren.

Environment Canada: 2000, "Strategic Environment Assessment at Environmental Canada". *Environmental Assessment Branch, Hull, Canada.*

Flood, J.: 1997, "Urban and housing indicators" , *Urban Studies,* **34,** 1635--1666.

Griliches, Z.: 1971, "Hedonic Price Indexes of Automobiles: An Econometric Analysis of Quality Change" in Zvi Griliches (ed.), *Price Indexes and Quality Change*, Cambridge: Cambridge University Press.

Jiliberto, R.: 2002, "Decisional environment values as the object of analysis for strategic environmental assessment", *Impact Assessment and Project Appraisal*, **20**, 61--70 - Beech Tree Publishing, Surrey, U.K.

Leontief, W: 1951, *The structure of American economy, 1919-1939* - 2nd edition, Oxford University Press, New York.

McHarg, I.L.:1968, *Design with Nature.* Natural History Press, Garden City, N.Y.

MacLaren, V.: 1994, "Developing indicators of urban sustainability: A focus on the Canadian experience". *Intergovernmental Committee on Urban and Regional Research,* Environment Canada, Hull, Canada.

Meadows, D.L. In: Shunji Murai (ed): 1995, "Towards Global Planning of Sustainable Use of the Earth: Development of Global Eco-engineering".

Morris, P. and Therivel, R.: 1994, *Methods of Environmental Impact Assessmen,*. UBC Press, Vancouver.

Munier, N.: 2002, "Impact assessment with urban sustainable indicators". *Proceedings of the 22nd IAIA02 Conference, Den Haag, The Netherlands, International Association for Impact Assessment*, Fargo, ND, USA .

Nitti, R. *Reaching the poor through sustainable partnerships: The slum sanitation program in Mumbai, India*. The World Bank - Urban Notes, 2002.

Noble, B.: 2002, "Strategic environmental assessment of Canadian energy policy". *Impact Assessment and Project Appraisal.* **20**, 177--188 - Beech Tree Publishing, Surrey, U.K.

Noorbakhsh, F. and Ranjan, S: 1999, "A model for sustainable development: integrating environmental impact assessment and project planning". *Impact Assessment and Project Appraisal* **17**, 283--292 - Beech Tree Publishing, Surrey, U.K.

OECD.: 1995 "The economic appraisal of environmental projects and policies: A practical guide". OECD, Paris.

Ottevanger, W., Deimel, M., Spaander-van Gendt, K.: 2000, "Infrastructure planning - the environmental impact assessment for a Netherlands-Germany rail link". *Impact Assessment and Project Appraisal.* **18**, 77--85 - Beech Tree Publishing, Surrey, U.K.

Polèse, M.: 1994, *Economía regional y urbana,* Editorial Tecnológica de Costa Rica, Cartago, Costa Rica.

Rosen, S.: 1974, "Hedonic Prices and Implicit Markets: Product Differentiation in Pure Competition" - *Journal of Political Economy*, **82**, 34--55.

Ranasinghe, M.: 1998, "Thoughts on a methodology to analyze viability of private -sector participation in new infrastructure projects in developing countries", *Impact Assessment and Project Appraisal.* **16**, 203--213 - Beech Tree Publishing, Surrey, U.K.

Saaty, T.L.: 1990, *Multicriteria Decision Making - The Analytic Hierarchy Process*, Volume I, AHP Series, McGraw-Hill, New York, NY.

Seinfeld, J.H and Pandis: 1988, *Atmospheric Chemistry and Physics of Air Pollution*, J. Wiley & Sons.

Shannon, C.E.: 1948, "A mathematical theory of communication". *The Bell System Technical Journal,* **27**, 379--423.

Souza, R.T., Santos, R.F., Genovez, A.I.B., Genovez A.M.: 2000 "Environmental impacts resulting from construction of a pisciculture installation in Juquiá and region, Brazil". *Impact Assessment and project Appraisal.* **18**, 335--340 -Beech Tree Publishing, Surrey, U.K.

Stolp, A., Groen, W., van Vliet, J. and Vanclay, F.: 2002, "Citizen values assessment: incorporating citizen's value judgments in environmental impact assessment",. *Impact Assessment and Project Appraisal.* **20**, 11--23 - Beech Tree Publishing, Surrey, U.K.

Thérivel, R. and Minas, P.: 2002, "Ensuring effective sustainability appraisal", *Impact Assessment and Project Appraisal.* **20**, 81--91 - Beech Tree Publishing, Surrey, U.K.

The World Bank: 1999, *A strategic view of urban and local government issues: Implications for the Bank.*
Transportation, Water and Urban Development Department, Washington, D.C., USA.

The World Bank: 1997, "Expanding the measure of wealth - Indicators of environmentally sustainable development" - *Environmentally sustainable development studies and monographs series* No. **17**, 99--109, Washington, D.C., USA.

Warner, L.L. and Diab, R.D.: 2002, "Use of geographic information systems in an environmental impact assessment of an overhead power line". *Impact Assessment and Project Appraisal.* **20,** 39--47 - Beech Tree Publishing, Surrey, U.K.

Wende, W.: 2002,"Evaluation of the effectiveness and quality of environmental impact assessment in the Federal Republic of Germany". *Impact Assessment and Project Appraisal.* **29,** 93--99 - Beech Tree Publishing, Surrey, U.K.

World Commision on Environmernt and Development: 1987, *The Bruntland Report also known as Our Common Future.* Oxford University Press, U.K.

Zeleny, M : 2000, *New frontiers of decision making for the information technology era.* Published by Yong Shi (University of Nebraska, Omaha) & Milan Zeleny (Fordham University, New York).

INDEX

AHP 5, 52, 82, 133, 145, 148, 156, 159, 187, 188, 189, 245, 251, 252, 298, 307
Alternatives . 3, 4, 5, 8, 9, 11, 12, 18, 21, 27, 29, 30, 45, 51, 53, 57, 61, 69, 84, 89, 101, 114, 132, 144, 150, 156, 161, 168, 177, 190, 200, 202, 251, 253, 256, 266
Analytical Hierarchy Process......7, 47, 52, 132, 145, 159, 172, 177, 188, 245, 251
Apparent impact, ...36
Assessment of significance........56
Batelle.......................................98
Carrying capacity......................25
Checklists 25, 70, 75, 97, 296
Comparison of techniques........ 45, 95, 97
Construction of an LPG pipeline 170
Construction of an overhead power line 169
Contingent valuation.............. 5, 23, 98, 104, 105, 108
Correlation analysis........ 158, 237, 243
Cost-Benefit analysis............. ... 5, 23, 27, 28, 106, 107, 175, 176, 191, 207, 270, 273, 276
Cost-Effectiveness analysis.....5, 28, 114, 190, 191, 276 ....
Criteria.................................... 3, 4, 5, 6, 22, 30, 31, 32, 33, 44, 47, 48, 132, 136, 140, 146, 147, 169, 171, 172, 179, 208, 212, 217, 251, 253
Criteria relationships 59
Cumulative effects.... ....25, 37, 59, 65, 103, 118, 144, 232
Delphi method......................... ... 61, 90, 91, 92, 97, 98, 175
Dollar value appraisal..............134
Dollar value with monetary restrictions 139
Dose-response function.......... ... 92, 93, 158, 296
EIA 3, 4, 5, 6, 7, 8, 9, 11, 12, 13, 14, 16, 17, 18, 19, 21, 22, 23, 24, 25, 26, 29, 31, 32, 33, 45, 57, 58, 59, 60, 61, 66, 69, 91, 98, 99, 100, 106, 121, 132, 133, 134, 154, 157, 158, 159, 160, 167, 170, 173, 177, 190, 191, 195, 197, 198, 200, 202, 205, 207, 209, 231, 232, 237, 245, 251, 253, 261, 265, 267, 268, 270, 273, 279, 280, 282, 299, 301, 285
EIA and the city.......................198
Eigen analysis..........................268
Environmental criteria..............98
Environmental damage appraisal 139
Environmental impact assessment 8, 25, 26, 45, 59, 66, 101, 169, 170, 190, 253, 271, 273, 275, 279, 307, 308
Environmental impact statement 9, 273, 274
Factor analysis.... ....47, 56, 57, 58, 237, 241, 245, 251, 265
Flow diagram.....................87, 127
Geographic Information System. 5, 25, 89, 99, 169, 176, 302, 308
Goal Programming............. ... 133, 159, 167
Hedonic pricing 109
Impact definition 34
Impact mitigation 61
Impacts 3, 4, 5,7, 8, 9, 11, 12, 19, 21, 23, 25, 27, 34, 40, 42, 44, 45, 55, 59, 63,

65, 70, 77, 78, 92, 95, 97, 101, 106, 110, 122, 133, 197, 200, 205, 231, 269, 272, 273, 280, 308
Input-output analysis 5, 117, 118, 120, 122, 124, 177, 188, 189, 192
Interaction between impacts 42, 43, 291
Internet references for Appendix... 268
Internet references for Chapter 1 23
Internet references for Chapter 2 44
Internet references for Chapter 3 64
Internet references for Chapter 4 97
Internet references for Chapter 5 187
Internet references for Chapter 6 232
I-O analysis...............................7
Life cycle analysis.................... . 5, 118, 122, 145, 189, 190, 193, 261
Likelihood of impacts.... ....39, 43, 44
Linear Programming...........................59, 160,188, 189, 253
Linear transformation 7, 237, 247, 269, 301
LPG pipeline.......................... 170
Mathematical Programming ... 132, 133, 139, 156, 159, 166, 167, 202, 214, 219, 220, 229, 259, 260, 267, 268
Matrices.................................. ... 7, 25, 94, 145, 159, 237, 244, 245, 296
McHarg systems.................. .... 87, 296, 306
Monitoring.................................7, 8, 9, 23, 25, 41, 55, 189, 209, 216, 231, 232, 273
MP 5, 132, 133, 139, 156, 159, 166, 167, 202, 214, 219, 220, 245, 259, 260, 267, 268, 298
Multicriteria analysis...............5, 6, 7, 15, 49, 85, 132, 134, 145, 159, 167, 187, 200, 207, 214, 224, 225, 253
Opportunity cost.... ....112, 159, 160, 167, 266, 267
Options to alleviate traffic congestion (The Netherlands).. 169
Overlays.................................. 87, 88, 89, 97, 100
Performance matrix.. ....136, 137, 138, 152, 161, 164, 254, 297
Population.. ....8, 9, 12, 13, 15, 25, 26, 34, 41, 43, 44, 45, 49, 60, 61, 70, 74, 97, 101, 135, 161, 163, 164, 168, 170, 176, 188, 190, 212, 225, 253, 269, 274, 275, 291, 302, 285, 286, 287, 288
Population impact.....................41
Positive impact...................... ..78, 79, 80, 81, 82
Public opinion.... ....31, 32, 158,180,186,270
Public participation 4, 6, 61, 169, 270
Regression analysis 7, 108, 237, 243, 265, 268
Remediation.... ....9, 12, 18, 20, 37, 39, 60, 74, 160, 253, 302
Residual effects...................... ... 9, 41, 43, 56, 291
Residual impact. 37, 41, 43, 290
Restoration.. ....18, 21, 37, 60, 61, 215, 216, 303

Risk analysis............................6, 22, 23, 40, 49, 54, 162, 164, 165, 172, 173, 174, 177, 191, 271, 303
Routing for an oil pipeline....... 170
Scoping.... ....24, 25, 31, 33, 52, 59, 65, 66, 160, 172, 190, 272, 305
Screening.... ....24, 25, 59, 63, 271
Selection of a sewage system .. 168
Selection of urban indicators ... 224
Sensitivity analysis 90, 167, 175, 267, 268
Sewage...... ....111, 206, 286
Shadow prices.... ....28, 53, 107, 108, 110, 160, 167, 176, 267
Slum upgrading 225
Social 4, 6, 11, 12, 13, 16, 17, 18, 22, 27, 28, 29, 30, 34, 49, 50, 54, 58, 59, 61, 62, 69, 70, 73, 78, 80, 90, 91, 106, 134, 154, 155, 156, 160, 161, 162, 164, 165, 169, 170, 171, 176, 202, 207, 208, 210, 212, 215, 216, 217, 218, 219, 224, 226, 257, 270, 275, 283, 292, 305
Spatially related impact 38
Stepped or chained matrices...... 94
Strategic environmental assessment.......... ....171, 272, 287
Sustainability 4, 7, 14, 15, 16, 18, 32, 53, 70, 78, 81, 132, 168, 207, 208, 209, 214, 217, 219, 220, 224, 232, 270, 275
Sustainable indicators............. 164, 214, 219, 224, 303, 306
The baseline. 33,39, 103, 115
Thresholds............................ ... 23, 43, 47, 48, 49, 53, 54, 55, 61, 65, 85, 103, 133, 143, 144, 163, 164, 166, 167, 180, 201, 207, 209, 217, 219, 229, 231, 233, 255, 270, 280, 295, 297
Thresholds standards 55, 143
Tools for impact identification .. 70
Urban and regional projects.....205
Waste incinerators 9, 38, 89, 100, 168,269

9 781402 020896